A "Hands On" approach to teaching . . .

Number and Operations

Grades 3 - 8

Scott Purdy
Sharon Rodgers
Linda Sue Brisby
Andy Heidemann
Natalie Hernandez
Jeanette Lenger
Ron Long
Petti Pfau

HANDS ON, INC.
SOLVANG, CALIFORNIA

Layout and Graphics: Scott Purdy
Cover Art: Petti Pfau
Illustrations: Petti Pfau

Order Number: HO 107 - I (Intermediate - Grades 3 - 8)
ISBN 0 - 927726 - 07 - 6

HANDS ON, INC.
2121 Rebild Drive
Solvang, CA 93463

Introduction

This is the seventh book in our series aimed at helping teachers teach mathematics more effectively through the use of manipulatives. Each of our books **Statistics, Probability, and Graphing**, **Measurement**, **Logic**, **Geometry**, **Algebra**, **Pattern and Function**, and **Number and Operations** focuses upon a math"strand" as identified in the California State Mathematics Framework. There is also a close match between our books and the NCTM Standards and Criteria for Mathematics Instruction.

We have found in our own classrooms that this "Hands On" approach has raised the understanding, performance, and motivation of our students.

As with our other books, we have included a Task Analysis which explains the various skills which are essential to a mastery of Number and Operations. Each skill, or Task, has from two to six manipulative activities which use common, inexpensive items -- many of which can be found in the classroom or at home. You will also note that the Task Analysis lists primary objectives and well as those for grades 3 - 8. Activities covering the primary objectives are presented in our **Number and Operations - Primary** book. We include the objectives in this text only to serve as a point of reference.

The Tasks Analysis is arranged in a hierachy by complexity and by grade level. We have included third through eighth grade ideas within one book because we find that it is convenient to have remedial, grade level, and enrichment activities in one source. This is particularly true in the area of number and operations. For example, when undertaking a study of rounding decimals, some students may be unclear as to how to round whole numbers while other students may be able to easily round decimals and need the challenge of intermixing fractions into this study. We provide activities which will appeal to each group, the average student, the remedial learner and the enrichment student.

We recognize that the majority of math textbooks dedicate at least 75% of their contents to number and operations. We do not provide worksheet pages in our text nor do we stress teaching the algorithms for math computation. And while we are 100% committed to students learning the algorithms, we want students to understand WHY the algorithms work. We've presented some unusual and unique approaches to this purpose.

Hands On, Inc. is comprised of eight teachers at Solvang Elementary School in Solvang California. We continue to teach as we write and publish these texts.

We have included 100 lessons on number and operations that will work for you in the classroom. We think you will enjoy sharing these lesson ideas with your class.

Number and Operations
Task Analysis

Primary

Numbers	P1. Identifies, names, and write numbers.	Primary
Numbers	P2. Identifies that numbers represent a quantity.	Primary
Numbers	P3. Places numbers into a sequence and compares number values using greater than, less than, equal to, and even.	Primary
Counting	P4. Counts by twos, threes, fives, and tens.	Primary
Place Value	P5. Identifies place value.	Primary
Addition/Subtraction	P6. Understands and uses basic addition and subtraction facts through ten.	Primary
Addition/Subtraction	P7. Understands and uses the terms of addition and subtraction (plus, addend, difference, less than, minus, joining sets, removing subsets).	Primary
Rounding	P8. Rounds numbers to the nearest 10, 100, and 1000.	Primary
Addition/Subtraction	P9. Uses estimation to mentally solve addition and subtraction problems.	Primary
Fractions	P10. Identifies and uses the fractions 1/2, 1/3, 1/4, 2/3, and 3/4.	Primary
Expanded Notation	P11. Writes numbers in expanded notation and translates verbal numbers to digits.	Primary

Grades 3 - 8

Place value	1. Identifies the place value and writes whole numbers through millions.	Middle
Expanded notation	2. Writes numbers in expanded and scientific notation.	Middle
Rounding	3. Rounds off numbers.	Middle
Place value	4. Identifies the place value and writes decimal numbers through thousandths.	Middle/Upper
Addition/Subtraction	5. Computes addition and subtraction using whole numbers and decimal numbers.	Middle/Upper

Multiplication	6. Identifies and uses the basic elements of multiplication including basic facts, terminology, prime numbers, composite numbers, and factors.	Middle/Upper
Multiplication	7. Represents multiplication in arrays and as repeated addition.	Middle/Upper
Multiplication	8. Uses various methods to estimate products.	Middle/Upper
Multiplication	9. Performs multiplication computation including those that require regrouping.	Middle/Upper
Division	10. Identifies and uses the basic elements of division including basic facts, terminology, and remainders.	Middle/Upper
Inverse operations	11. Identifies addition/subtraction and multiplication/division as inverse operations.	Middle/Upper
Division	12. Uses various methods to estimate quotients.	Middle/Upper
Division	13. Performs division computations.	Middle/Upper
Problem formats	14. Identifies and writes addition, subtraction, multiplication, and division problems in vertical or horizontal format.	Middle/Upper
Fractions	15. Understands and uses the terminology of fractions.	Middle/Upper
Fractions	16. Understands ordering and rounding off of fractions.	Middle/Upper
Fractions	17. Reduces fractions and identifies equivalent fractions.	Middle/Upper
Fractions	18. Adds and subtracts fractions with like and unlike denominators.	Middle/Upper
Fractions	19. Multiplies and divides fractions, mixed numbers, and whole numbers.	Middle/Upper
Decimals	20. Multiplies decimals.	Upper
Decimals	21. Divides decimals and computes and rounds remainders.	Upper
Decimals	22. Identifies equivalent fractions, decimals, and percents	Upper
Percent	23. Estimates percentages.	Upper
Percent	24. Computes percent of a number.	Upper
Percent	25. Computes percentage relationships between two numbers.	Upper
Proportion/Percent	26. Computes rates, ratios, proportion, scale, and percentage.	Upper

Table of Contents

1	Identifies the Place Value and Writes Whole Numbers Through Millions

Stand and Deliver

Grade Level: Middle

MATERIALS: A card for each student with a number 0 through 9 written on both sides

ORGANIZATION: Teams of ten students

PROCEDURE: In this activity, students will be working together to physically represent a large number generated by another team.

Begin by dividing the class into two or three teams of ten students. If there are extra students, they can be given special tasks such as "number reader," "number checker," and score-keeper.

Each student should be given a card and the digits 0 through 9 should be written on the cards -- each team member should have a different digit. Each student should then write a number (the teacher should decide if the numbers should include millions, ten millions, etc.) which is to be read aloud for the class. The number must not repeat any digits but need not include all digits.

The game begins with a member of team one reading a number aloud (i.e. three million six hundred four thousand nine hundred twelve -- note that only one zero is used). At this point, members of team two should come forward and place themselves in the proper arrangement to display this number to the class; each student finding the proper place for the number written on his or her card.

If the number is done correctly, the team gets five points. If the number is incorrect, it is re-read and the team may reorganize. If correct on the second try, they get three points. The third try would be worth only one point. Reverse roles with a member of team two becoming the reader for the second number.

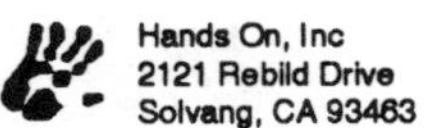

Hands On, Inc
2121 Rebild Drive
Solvang, CA 93463

1	**Identifies the Place Value and Writes Whole Numbers Through Millions**

Odometers to Go

Grade Level: Middle

MATERIALS: Cardboard rolls from paper towels or toilet paper, colored construction paper, markers, tape, scissors

ORGANIZATION: Individually or teams of two

PROCEDURE: The object of this lesson is to make an odometer which students can use to practice place value of large numbers.

Begin by having students cut strips of different colored paper. Each strip should be about 2 cm. in width and long enough to wrap around the paper roll. Use a different color for each place value (red for ones, blue for tens, etc.).

After strips are cut, students should wrap them around the roll and make a small mark where ends meet. They should number each strip 0 - 9 making certain that they space numbers properly. This, in itself, is an exercise in measurement.

Once numbers have been printed, wrap each strip around the roll. Tape the ends together. Strips should be loose enough to slide as they are rotated, yet not so loose that they move without being turned.

Once odometers are complete, have students practice using them by setting each number to zero and then, as they walk around the room or playground, turning a number for each step. Remind students that each time a number gets to "nine" in one column and it is turned to zero, then the number in the next column to the left must be increased by one.

Hands On, Inc
2121 Rebild Drive
Solvang, CA 93463

1	**Identifies the Place Value and Writes Whole Numbers Through Millions**

7 Up

Grade Level: Middle

MATERIALS: Game cards (Appendix A), teacher-prepared number cards, beans

ORGANIZATION: Whole class activity

PROCEDURE: This activity uses a Bingo type game to practice place value.

Give each student a game card. Have them randomly write in numbers (0 - 9) to fill each square (each child's card should be different). Explain that they should place a bean on each square that has an equivalent place value as the number card which is held up by the teacher.

For example, if the teacher were to hold up the card: Three hundred nine thousand, four hundred seventy-two, students should place a bean on a 3 in the hundred thousands place, a bean on 0 in the the ten thousands place, one on the 9 in the thousands place, etc.

When a child has seven beans in a row in a horizontal, diagonal or vertical pattern, he/she yells "Seven up." Numbers are checked by having the student read aloud the placement of each bean. For example, "a 3 in the hundred thousands place." "A 0 in the ten thousands place," etc.

millions	hundred thousands	ten thousands	thousands	hundreds	tens	ones
3	6	2	0	4	9	1
7	4	6	5	8	0	3
1	8	5	4	2	1	5
2	1	9	Free	6	7	0
9	9	3	6	8	8	7
8	3	1	9	0	4	2
5	2	4	7	7	6	9

Hands On, Inc
2121 Rebild Drive
Solvang, CA 93463

1	Identifies the Place Value and Writes Whole Numbers Through Millions

When the Chips Are Down

Grade Level: Middle

MATERIALS: Charts as shown below, lists of number words (Appendix B), counters such as beans or paperclips

ORGANIZATION: Whole class for explanation and then individually

PROCEDURE: This activity gives students practice in breaking down large numbers by place value. Each student should have a column chart similar to that shown below (Appendix B). You may want to start with smaller numbers and work up to millions and billions.

Begin by writing an easy number on the board (i.e. one thousand, three hundred, four). Have students fill in the spaces in the proper columns with counters (one counter to one space) to show the number on their chart. You will need to demonstrate this by writing it on the chalkboard or overhead after you check for the students' accuracy.

Point out that the tens column has no markers in this example and ask students what they would place there if there were to actually write the number. Also ask why they would place a zero in the tens place.

Continue this process with other, more difficult numbers involving zeros. You might also point out that each column on their chart has only nine spaces for markers. Ask why there are only nine.

Hundred Billions	Ten Billions	Billions	Hundred Millions	Ten Millions	Millions	Hundred Thousands	Ten Thousands	Thousands	Hundreds	Tens	Ones

Hands On, Inc
2121 Rebild Drive
Solvang, CA 93463

2	Writes Numbers in Expanded and Scientific Notation

Character Reference
Grade Level: Middle/Upper

MATERIALS: Construction paper, newspaper, glue, and scissors

ORGANIZATION: Individually or in groups of two or three

PROCEDURE: This is a lesson in expanded notation which gives students a graphic display of relationship of place value.

Write the number 2,000 + 300 + 60 + 7 = 2,367 on the chalkboard and explain that students will be making a display of this number by using a sampling count of the number of letters in a newspaper column.

Students will need to be shown how to do the sampling. They should count the number of letters in several lines of a newspaper story and estimate the average number of letters per line. They should then multiply this number to decide how many lines 1000 characters would fill. In our experience, there are approximately 28 characters per line in our local paper. Therefore, approximately 35 and 1/2 lines will contain 1000 letters.

Students should cut out two sets of 1000 characters, three sets of 100 characters, six sets of 10 characters, and seven sets of 1 character to represent the given number in expanded notation.

Have students then prepare their own representation of a number they have created. on a piece of construction paper. They should not write the actual number on their construction paper representation. Instead, have them display their newspaper sets to the class and have the class figure the number.

Students may choose to work with ten thousands or even hundred thousands. Let them experiment with these numbers. The visual representation is memorable for the class.

2	**Writes Numbers in Expanded and Scientific Notation**

Name That Number!

Grade Level: Middle/Upper

MATERIALS: No special materials are necessary

ORGANIZATION: Whole class activity

PROCEDURE: This activity uses a game show approach to give students practice in using expanded or scientific notation.

The teacher should model the first game by drawing seven empty squares on the chalkbaord. A large number should be written on a slip of paper and kept secret. This is the number that students will have twenty tries to guess. For our example, we will use the number 2,354,698. Call on one student to come forward and write a number in each of the boxes.

5	6	8	3	0	6	1

At this point, the teacher should select a student and give a clue, such as, "the number is off by 3 million." The student can then come forward and change the number in the box. The choices would be to increase it to 8 million or decrease it to 2 million. In this case, if the student wrote an 8, the teacher would give another clue, such as, "the number is off by 6 million."

Proceed in this manner until the students have successfully matched the number written on the slip of paper. When finished, call on several students to read the number in various ways (traditional, expanded, scientific). Select a student to take the place of the teacher in the second game.

If working on scientific notation, a sample clue might be, "the number is off by 3 plus 10 to the sixth power."

Hands On, Inc
2121 Rebild Drive
Solvang, CA 93463

2	Writes Numbers in Expanded and Scientific Notation

Pinning It Down

Grade Level: Middle/Upper

MATERIALS: String, two or three clothes pins for each student, index cards and markers

ORGANIZATION: Whole class activity

PROCEDURE: Prepare for this lesson by tying a string across the front of the classroom. This activity will give students practice in ordering numbers written in various formats.

Give each student two or three clothes pins and some index cards. Each student should write a whole number on the first card, a different number in expanded notation on the second card, and a third, different number in scientific notation on the third card (this may vary depending upon the level of the class). Students should be given a range to work with, for example, the numbers must fall between 1 and 10,000.

Discuss the idea of the string acting as a number line and have a class member come forward and place a clothes pin on a random point to represent "zero" (to establish a point of reference) on the line. At this point, select one student at a time to come forward and place a clothes pin (along with an index card) at the proper point on the string. Continue with four or five more students asking each to place their clothes pin and card on the appropriate position on the string.

As a class, discuss the placement of clothes pins thus far. Adjust those pins which are misplaced or need more "space" between numbers. In groups of five, have the remainder of the class come forward to place their first card/clothes pin. Have the students exchange their second and third card with one another and have them place these cards on the line. Discuss the placement when they have finished.

As an extension, you can use this activity to do comparisons of fractions and decimals.

2	Writes Numbers in Expanded and Scientific Notation

Place Value Shuffleboard

Grade Level: Middle/Upper

MATERIALS: Assortment of beans (five different kinds), paper (8.5 x 11), marking pens

ORGANIZATION: Teams of two or four

PROCEDURE: This activity will give students practice at identifying numbers written in traditional and in scientific notation. It is similar to shuffleboard and can be modified to encompass other skills.

Begin by explaining to your students that they are going to do an activity in which they will be required to correctly write and say a number which will be randomly generated. They will score points for correctly identifying the numbers, no points will be awarded for mistakes. They will be using beans which will represent a given number: pinto = 5; navy = 4; kidney = 3; black = 2; and lima = 1.

Each person will roll/slide a bean onto a triangular scoring grid as shown below. This can be made using regular notebook paper and marking pens; it may also help to close off the scoring end with books to stop the beans from going too far. Each section of the grid represents a place value from 1's to 100,000's expressed in either traditional or in scientific notation.

Students should take three beans at random from a bag and roll them into the scoring triangle. When a bean lies across a line students should use the higher number. They must then correctly record the number generated by the location of the three beans and say the number aloud. For example, if a student were to slide a lima bean on 1, a black bean on 10^3, and a pinto bean on 10^5, he/she would then say, "five hundred two thousand one." (Note that the word "AND" is not used -- this should be reserved for decimal numbers.) Then the next player takes a turn.

As an extension you can add a row representing decimal places.

Hands On, Inc
2121 Rebild Drive
Solvang, CA 93463

3	Rounds Off Numbers

Going 'Round the Hill

Grade Level: Middle

MATERIALS: Graph paper, four colors of plastic chips, bags

ORGANIZATION: Teams of two

PROCEDURE: This activity gives students a tangible and visual way to represent rounding off numbers to the nearest tens and hundreds place.

In this activity, instead of using real money, students will be using chips -- one color to represent pennies, another color for nickels, dimes, and quarters. They will select three or four chips from a grab bag, total the amount, and then round this number to the nearest 100 and the nearest 10.

Model the activity by cutting out a strip of graph paper 100 squares long (this may need to be taped together). Explain that this strip is like a road that goes over a hill. Fold the paper half at the 50th square and lay the "hill" on a flat surface.

Explain that when rounding-off, there is a cut-off point; numbers falling on the "up-slope" round down, while numbers on the "down-slope" round up. The fold at the 50th square represents the cut-off point when rounding-off to the nearest 100.

Have students make their own 100 strips with the fold at 50; then, working in teams of two, have students select a handful of chips, total the amount and find this point on the "hill." They should decide whether the number rounds up or down.

Next, have them work together to see how they could fold the paper to represent rounding to the nearest ten (an accordian fold is the most common discovery). Have students take turns selecting chips and then rounding their total to the nearest 100 and the nearest 10.

Hands On, Inc
2121 Rebild Drive
Solvang, CA 93463

3	Rounds Off Numbers

Chipping Away for Years
Grade Level: Middle

MATERIALS: Encyclopedia, poker chips

ORGANIZATION: Groups of two

PROCEDURE: Begin this activity by discussing the idea of an average life span. Once students understand the concept, tell them that they will be researching life spans and rounding these numbers.

Have each group make a list of the average life span of 5 to 10 animals such as a chimpanzee, elephant, giraffe, tiger, tortoise and humans. They can find this information in an encyclopedia.

Using chips to represent 10 years, they should make stacks with labels indicating which animal each stack stands for as well as the actual years each animal lives. Have each student explain the difference between the rounded number and the actual age.

As an extension, this can be used to represent ratios between/among life spans.

4	Identifies the Place Value and Writes Decimal Numbers Through Thousandths

The Point of Decimals

Grade Level: Middle/Upper

MATERIALS: Cards with decimal numbers, blank cards, markers, scissors

ORGANIZATION: Teams of three

PROCEDURE: Prepare one set of decimal flash cards with the decimals listed below.

Display the set of ten teacher-prepared cards to the class. Ask students to discuss the similarity of the numbers and read each number aloud. Select one student at a time to come forward and move one card with the ultimate goal of placing the cards in order from least to greatest.

Continue selecting students to come forward and make adjustments until the correct order is accomplished.

Next, tell students that they will be meeting in groups to prepare their own set of cards with the goal of trying to stump the other groups' efforts to place cards in order. Group members should use index cards to write their set of ten decimals. You may wish to limit the number of digits to specific place values.

Once cards are complete, have groups exchange cards and time one another to see how quickly the cards can be placed in numerical order. Each group should exchange card sets with every other group. Spot check the students' understanding by pointing to certain cards and have them read the numbers aloud.

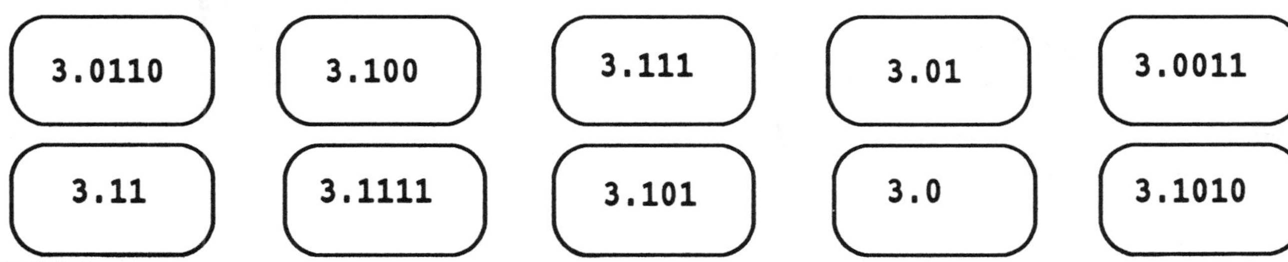

4	Identifies the Place Value and Writes Decimal Numbers Through Thousandths

How Sweet It Is

Grade Level: Middle/Upper

MATERIALS: Ten sugar cubes for each team, markers

ORGANIZATION: Teams of two students

PROCEDURE: This activity uses sugar cubes to generate random numbers for students to read aloud.

Divide the class into teams of two students and give each team ten sugar cubes and a marker. On each cube face, they should write one digit from 1 - 9 in any order they wish. On three of the cubes, one face should be marked with a decimal point.

Students should sit next to each other and place the cubes in a horizontal line on a desk. Player one begins by reading the number which appears on the top surface of the cubes. If the number is read correctly, player one receives a point and can turn any one cube for player two to read. If player one is incorrect, player two gets to read the number and receives a point; in this case, player two gets to rotate one of the other cubes.

The only rule which dictates which cube can be rotated is that ONLY ONE DECIMAL POINT MAY BE DISPLAYED AT ANY TIME.

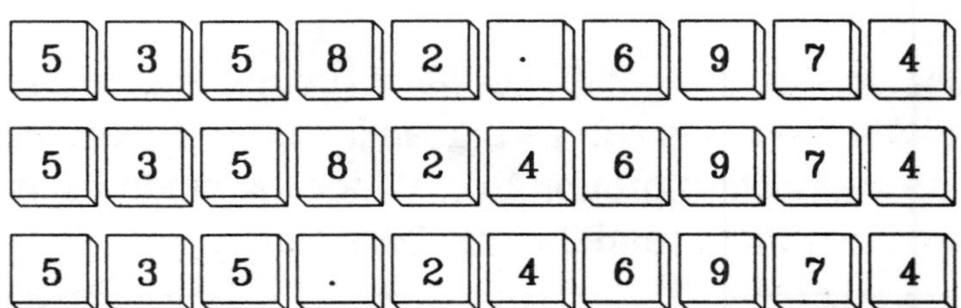

Give students time to complete a game to five points and then switch partners. If students have difficulty with ten numbers, you can reduce the number of sugar cubes accordingly.

Hands On, Inc
2121 Rebild Drive
Solvang, CA 93463

4	Identifies the Place Value and Writes Decimal Numbers Through Thousandths

Making a Point of Order

Grade Level: Upper

MATERIALS: Five 3" x 5" cards for each student, markers

ORGANIZATION: Individually or in teams of two if the concept is new to students

PROCEDURE: Placing whole numbers into numerical order is relatively simple for most upper grade students; however, decimal numbers make this much more confusing. This activity gives students an opportunity to practice this skill.

Give five cards to each student (or group) and tell them that they will be writing a different decimal number between 3 and 4 on each card. The only "rules" are that each number must have a decimal point and no number may have more than 6 digits. You might also mention that cards must be written neatly since they will be distributed to classmates.

Once students have finished writing, collect all cards and shuffle them thoroughly.

Regroup the cards into stacks of five and give a stack to each student or group. The students should take their five cards and arrange into numerical order from least to greatest. Have students check their work with a partner. The teacher will need to circulate through the room answering questions.

Once students correctly order the first cards, have them rotate card sets around the room.

3.2354	3. 24	3.44665	3.7	3.766

Hands On, Inc
2121 Rebild Drive
Solvang, CA 93463

4 | Identifies the Place Value and Writes Decimal Numbers Through Thousandths

Taking Orders

Grade Level: Middle/Upper

MATERIALS: Slips of paper and pencils

ORGANIZATION: Teams of two

PROCEDURE: This activity gives students practice in deciphering large decimal numbers as well as determining greater than and less than clues.

Divide students into teams of two and have each student write a large number extending from millions to ten-thousandths. Students should incorporate zeros into their examples. A sample number might be: 3,009,806.5004. The number should be kept secret from the team partner.

Using several slips of paper, every word the student would say in reading this number should be written on a separate slip of paper. In the case of the example given, the set of numbers would be:

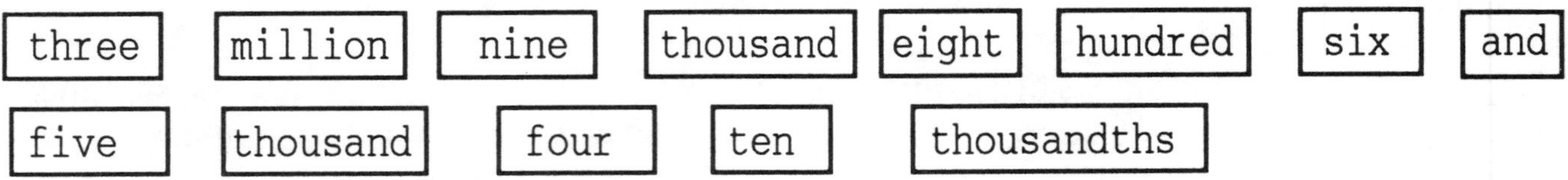

Each student should then shuffle their set of cards and exchange with their partner. Students should take turns trying to place the cards in the correct order. The only clues that should be given are "greater than" and "less than." As each clue is given, the student placing the cards should move numbers around until eventually the set of cards matches the number originally written by his/her partner.

5	Computes Addition and Subtraction Problems Using Whole Numbers and Decimal Numbers

Bundles of Fun!

Grade Level: Middle

MATERIALS: Soda straws and rubber bands

ORGANIZATION: Individually or in teams of two

PROCEDURE: In this activity, students will use bundles of straws to practice regrouping in subtraction.

Have students cut the straws into approximately 2 inch (5 cm) lengths. Using rubber bands, have students make 19 bundles of ten straws. Using ten of these bundles, make one bundle of 100 straws. Students will manipulate these bundles to perform subtraction regrouping.

Begin by demonstrating a simple problem such as 26 — 18. Lay out the straws to show 26 (2 bundles and 6 single straws) and 18 (one bundle plus eight singles). Ask the students how you could subtract 8 from 6. Someone will reply that you have to unwrap one bundle in order to do this.

Have students remove the rubber band from one bundle and combine the 10 straws with the 6 singles. Now ask if it is possible to do the subtraction. Have students show the difference with other bundles of straws and have them then write the algorithm for this example.

Have students work in groups or individually to repeat this procedure with other subtraction examples written on the chalkboard.

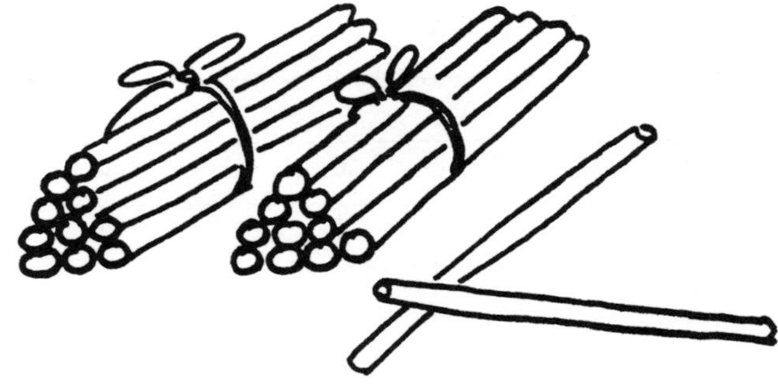

5	Computes Addition and Subtraction Problems Using Whole Numbers and Decimal Numbers

Have a Few Coins

Grade Level: Middle

MATERIALS: Play money, chart paper, pens

ORGANIZATION: Teams of two or more

PROCEDURE: This activity gives students practice in finding different combinations of addends to total a given sum along with practice in totaling money amounts.

The teacher should begin by writing an arbitrary figure on the board such as $.37. Ask students to suggest different combinations of coins which would total this amount. Record student responses on a chart as shown below.

Have students decide which combination on the list has the fewest and which has the most coins. Tell students that it will be their task to use play money to show the fewest number of coins that can be used to total amounts suggested by classmates.

Distribute play money to each group and select one child at a time to say a money amount. Have group members work together to display this amount.

Dollars	Half-dollars	Quarters	Dimes	Nickels	Pennies
0	0	1	1	0	2
0	0	0	3	1	2
0	0	0	0	7	2
0	0	1	0	3	2

Hands On, Inc
2121 Rebild Drive
Solvang, CA 93463

5	Computes Addition and Subtraction Problems Using Whole Numbers and Decimal Numbers

Tuning in on Decimals

Grade Level: Middle/Upper

MATERIALS: AM/FM radio with digital display large enough for students to see

ORGANIZATION: Whole class activity

PROCEDURE: This is a "just for fun" activity in which students tune through the radio dial identifying the decimal frequency of various radio stations. They then use this data to perform different decimal addition and subtraction calculations.

Begin by discussing the way in which radio stations are different from one another. Question students as to why stations don't overlap sound on one another (because each station has its own frequency). Tune in one of the local radio stations and let students see that it has a specific station number, for example, 94.7.

Next have students list the various types of music played by stations. They might include country-western, rock, easy listening, classical, etc.

Begin at one end of the tuner and have students record the readout for every station and the type of music they hear. From this list, have them do various computation such as: the sum of their favorite stations, the sum of all country-western stations, the difference between rock and roll stations and country-western stations, or the difference between the total of all AM stations vs. FM stations.

As an extension, the information can be nicely presented in the form of a graph.

Hands On, Inc
2121 Rebild Drive
Solvang, CA 93463

5	**Computes Addition and Subtraction Problems Using Whole Numbers and Decimal Numbers**

Graphic Decimals

Grade Level: Middle/Upper

MATERIALS: Graph paper, crayons/markers, playing cards with 10's through kings removed

ORGANIZATION: Individually or in pairs

PROCEDURE: This is an activity to help students visualize how to subtract decimal numbers (i.e., remembering to "line up" the decimal points).

Explain that a decimal point can be viewed as a "point" of reference. It marks the place where whole numbers are separated from partial numbers (fractional numbers). Numbers to the left of the decimal are whole, numbers to the right are fractions. For this reason, when subtracting mixed numbers, students must align the decimal points -- so they subtract wholes from wholes and parts from parts.

Distribute graph paper and markers and have students draw a line down the center of the paper. Each student should generate two, 3 digit numbers by randomly selecting six cards from a deck. Have them lay the cards as shown to create a subtraction problem.

Whole numbers	Decimal numbers
3 4	1
2 2	4

Borrow

Students should now color in the number of squares represented by the top number (34 squares on the left side of the line and 1 square on the right side). From these squares, they should subtract (cross out colored squares to represent the bottom number) the amount underneath. The challenge the students will face in the given example is how to "borrow" from whole numbers to decimal numbers (for each square borrowed/taken from the left side, they may add ten to the right). Give teams time to discuss this and decide upon a solution. Have students share solutions with the class.

For those students who do not create a problem requiring regrouping, the process is very simple. The squares which are not crossed out become the answer. Encourage these students to reset their cards to solve a regrouping situation. Upon completion, emphasize the point that the decimal point is the reference point and therefore must be aligned when doing subtraction.

Hands On, Inc
2121 Rebild Drive
Solvang, CA 93463

5	Computes Addition and Subtraction Problems Using Whole Numbers and Decimal Numbers

Sum Will Get It!

Grade Level: Middle/Upper

MATERIALS: Numbers 0 - 9 cut out from construction paper

ORGANIZATION: Individually or in teams of two

PROCEDURE: This lesson combines logic with addition of decimals. You may want students to use calculators to check their answers; however, paper and pencil will work just as well.

Have each student or team cut numbers 0 - 9 from construction paper. They should cut out three or four decimal points as well. The goal of the activity is to have students arrange the numbers in such a way as to total a number (sum) given by the teacher.

There are three rules which students must follow: 1) they may not use any digit more than once, 2) each digit must be used each time, and 3) each set of addends must have a decimal fraction.

Once each student has a set of numbers, decimal points, and understands the rules, the teacher should call out a sum. Students then have three to five minutes to arrange their numbers to get as close to the sum as possible. In our example, we have used 8.4 as the target number/sum.

```
  1.8
   .59                      7.219
  2.40       8.4321          .056
+ 3.76     +  .5679       +  .834
  8.5        9.0000         8.109
```

5	Computes Addition and Subtraction Problems Using Whole Numbers and Decimal Numbers

An Eggcellent Example

Grade Level: Middle

MATERIALS: Egg cartons, beans or beads, markers, 3 x 5 cards

ORGANIZATION: Teams of two

PROCEDURE: Students are so familiar with base ten (decimal) that they don't question or notice how it works. In this activity students will learn how to borrow and regroup in various bases. We will use the example of base four for the purposes of explanation.

Cut the egg cartons into three sections and give one section to each team of students. Have them number the cups in their egg carton 1, 2, and 3. Also give them four 3 x 5 cards to be numbered the same way. Hand out a pile of beans to each team.

Begin by explaining how the decimal system works: the range of numbers is from 0 to 9 and when an amount goes beyond nine, it moves to the next place and a 1 is used before each successive number.

In base four only 0 through 3 exist. The carton represents the first place value or base (the total amount used before they start over again at the next "level"). After they have filled in the first level, what happens? Ask your class to discover how the value of 6 would be represented in base four. Have them place beans in each cup starting with 1 until they have filled the three sections. Once this "level" is full, they should place the first 3 x 5 card next to the carton and label it "1" to signify that one "level" or base is full. They should now see that 4 = 10 (pronounced one, zero) in base four. Have them proceed with their investigation to find out what six is (6 = 12). Pose several other numbers for them to represent in base four.

Once students understand the concept, have them work together to solve some simple addition and subtraction problems in base four. Call on students to describe aloud the process they are using as they regroup. Also discuss with students why they use only three egg sections for base four and why they would use only nine egg sections for base ten. Have students experiment with other base systems.

Hands On, Inc
2121 Rebild Drive
Solvang, CA 93463

6	Identifies and Uses the Basic Elements of Multiplication Including Basic Facts, Terminology, Prime Numbers, Composite Numbers, and Factors

By Any Other Name

Grade Level: Middle

MATERIALS: One die per group, beans, small scoop or spoon, paper plates, half a deck of cards per group

ORGANIZATION: Groups of two

PROCEDURE: Students should understand that mathematics has its own language. In this activity, they will develop an understanding of the "language" of multiplication.

Explain that each team of students is going to become a miniature shipping department at their desks. They will be responsible for shipping packages of certain raw materials. Distribute the materials described above to each pair of students.

Have students set up their factory in a production-line format starting with their raw materials (beans). Explain that the raw materials are called the multiplicAND in math since they AND something else are multiplied together. Another way to explain this is to have students imagine that the multiplicand is a "set" or "package" of a specified number of items.

The die is the multipliER since it is what the multiplicand is multiplied by (the idea of -er and -or endings meaning "one who does or works at something" could be explained by examples such as actOR or fieldER). The multiplier tells how many "sets" are to be shipped.

They should draw one card at random, this is the amount of raw material they will need to make one packaged set. Next, they should roll the die, this represents the number of packages to be shipped in the order. The multiplicand times the multiplier equals the amount of PRODUCT to be shipped. The product should be placed on the plate.

As students select cards and roll the die, have them write the problems they are solving by "filling the order." For example, if a student draws a ten card and rolls a 3, ten sets of three beans should be placed on the product plate. Students should then write 10 x 3 = 30 on a sheet of scratch paper.

6	Identifies and Uses the Basic Elements of Multiplication Including Basic Facts, Terminology, Prime Numbers, Composite Numbers, and Factors

Number Barbecue

Grade Level: Middle

MATERIALS: Straight macaroni, dice, paper clips which are straightened (to be used as skewers), a watch with a second hand

ORGANIZATION: Teams of two

PROCEDURE: In this activity your students will have a chance to practice their basic multiplication skills. It is very similar to the game of jack-straws or pick-up-sticks.

Begin by giving a large handful of macaroni, a die, and several skewers (paper clips) to each group. Tell them that they are going to have a timed race. Each player will roll a die to determine their target number (factor) and will then be given 60 seconds to accumulate as many sets of macaroni as possible.

The "catch" is that they can only use the skewer (no fingers, etc.) to collect the macaroni, and they must collect a full set at a time. For example, if the number rolled is five, the student must "skewer" five macaroni pieces. The first skewer is then set aside and a second skewer is used to collect another set of five. The player may not start a pile of three and then add two, nor may he/she skewer seven and return or save two.

After time is up, the player counts the number of piles and calculates the total number of sets of macaroni collected. The piles are then lumped together and the total number of macaroni counted to check the calculation. For example: after a five is rolled on the die, a student might collect three sets of macaroni. The student would then multiplies 5 times 3 and would receive 15 points for that round.

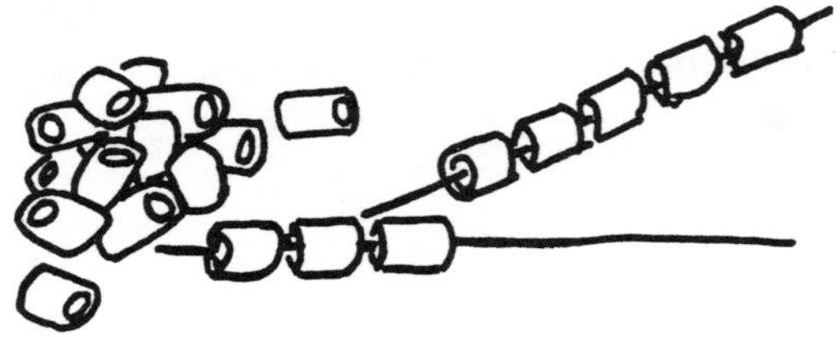

Students should continue adding points to their total with each round. The first student to 100 points is the winner.

Hands On, Inc
2121 Rebild Drive
Solvang, CA 93463

6	Identifies and Uses the Basic Elements of Multiplication Including Basic Facts, Terminology, Prime Numbers, Composite Numbers, and Factors

Prime Time Take Away

Grade Level: Middle/Upper

MATERIALS: Two dice for each team, stickers to renumber the dice, pencil, paper

ORGANIZATION: Teams of two

PROCEDURE: Tell students that in this activity, they will need to identify prime numbers. They will each start with an imaginary 100 points and must then subtract prime numbers from their 100 points to arrive at zero.

Divide students into teams of two and provide two number cubes for each team. Using small stickers, the students should label one cube with the digits 0, 1, 2, 3, 4, and 5. The second cube should be labeled 4, 5, 6, 7, 8, and 9. Students should roll the cubes and use the two digits in either order or individually. For example, if 2 and 3 were rolled, the student could make 23 or 32, or the student might choose to use just the 2 or the 3.

They are allowed to subtract prime numbers only. If no prime number can be made from the roll, the player loses his/her turn. If a player tries to subtract a number that is not prime, then 25 points are added to his/her score.

Players should keep track of their own scores. To win, they must subtract exactly to zero. If they roll a number which would take them to less than zero, they lose their turn.

6	Identifies and Uses the Basic Elements of Multiplication Including Basic Facts, Terminology, Prime Numbers, Composite Numbers, and Factors

Primarily Primes

Grade Level: Upper

MATERIALS: 3 by 5 cards, markers

ORGANIZATION: Pairs of students

PROCEDURE: Prime factorizations provide an interesting use of prime numbers which students often miss. This activity uses prime factorizations to show relationships between numbers.

Begin by writing 2 x 2, 3 x 2, and 3 x 3 on the chalkboard. Have students respond with the products and ask if there is anything they notice about the factors written or the products (factors are all prime, products are all composite).

Share with students that the prime factorizations for the numbers 2 -10 are written as follows. Tell students that they are going to meet in pairs and use number cards to find patterns of prime factorizations.

2	prime
3	prime
4	2 x 2
5	prime
6	2 x 3
7	prime
8	2 x 2 x 2
9	3 x 3
10	2 x 5

Give teams of students a stack of index cards and have them write the first 10 prime numbers: 2, 3, 5, 7, 11, 13, 17, 19, 23, 29, using a different card for each number (they will have to make several 2, 3, and 5 cards). Using these cards as a means of displaying different combinations, have them experiment with combinations of prime factors to do the prime factorization for all numbers through 50.

When students have finished with this first portion of the lesson, ask students to verbalize their discoveries about the patterns that they found. There are many statements that students can make including how prime factorization can be used to find the GCF and LCM and how prime factors represent exponential numbers.

Hands On, Inc
2121 Rebild Drive
Solvang, CA 93463

7	**Represents Multiplication in Arrays and as Repeated Addition**

Hip, Hip Array for the Turn!

Grade Level: Middle

MATERIALS: Teacher prepared cards with various arrays, magazines, scissors, paste, construction paper

ORGANIZATION: Individually

PROCEDURE: This is an early lesson in helping students understand that an array can represent a multiplication product and that multiplication is associative.

Begin by displaying an array (as shown below) for the students. Ask for a volunteer to come forward and write a number sentence on the chalkboard which would represent the array. The student might write 4 x 6 = 24 or 6 + 6 + 6 + 6 = 24. Either response is correct. Explain that an array usually depicts a multiplication product. In this case, 4 x 6 = 24.

Have a student come forward and point out how the array depicts this number sentence. Now rotate the card 90 degrees (second card below). Ask a student to come forward and write this number sentence on the board 6 x 4 = 24. Explain to students that multiplication is similar to addition in that the order of the numbers does not matter, the answer is the same because you can associate the numbers in any order.

Distribute the materials to the students and have them make up their own arrays from similar items cut out from magazines. The student arrays should each depict a different multiplication fact. When students have finished, have them come forward and repeat the process of displaying an array for the class and then turning it 90 degrees to represent the associative property.

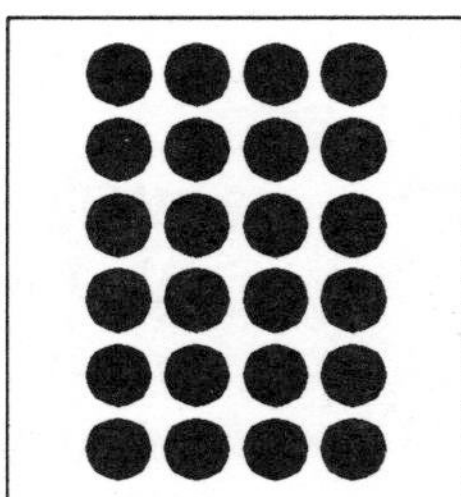

Hands On, Inc
2121 Rebild Drive
Solvang, CA 93463

7	Represents Multiplication in Arrays and as Repeated Addition

Taking Multiplication for a Spin

Grade Level: Middle

MATERIALS: Two spinners per group (Appendix C), pencil and paper

ORGANIZATION: Teams of two students

PROCEDURE: In this activity, students will work in pairs to draw arrays as shown below and then write the multiplication fact represented by the array.

Very little introduction to this lesson is needed as it is relatively simple for students to grasp. Divide the class into teams of two students, each student should have a spinner, and a sheet of paper and a pencil.

Students should simultaneously spin their two spinners. Suppose that the spinners have landed on 6 and 5. Students should look at one another's spinners and try to say the product of the two numbers; however, the first student should say the number of his/her spinner first, "6 times 5 is ..." The second student should say his/her number first, "5 times 6 is ..."

At this point, whether or not either student has correctly or incorrectly identified the product, they should both draw the array which represents the answer. From this, students should check their responses.

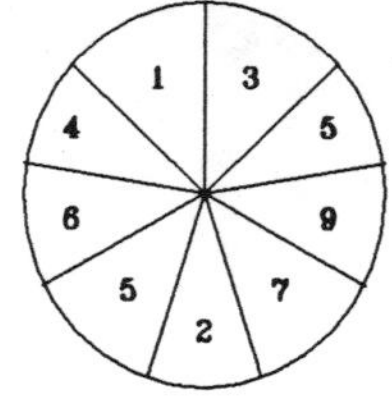

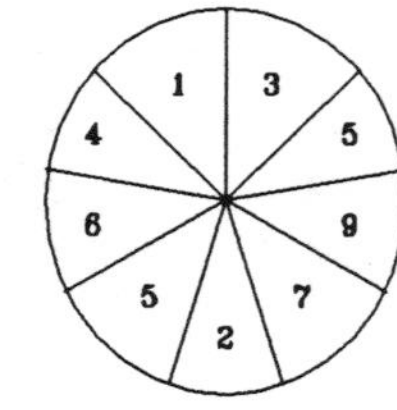

With time and practice, you may wish to make a game of this activity by awarding points for the first student to correctly identify the product or the first to correctly draw an array.

5 x 6 = 30

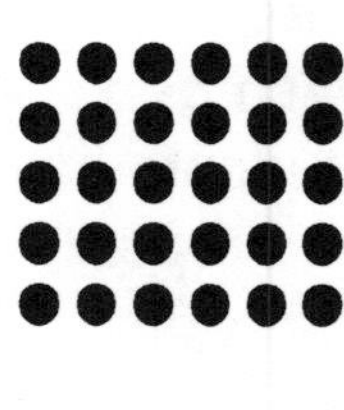

6 x 5 = 30

Hands On, Inc
2121 Rebild Drive
Solvang, CA 93463

7	Represents Multiplication in Arrays and as Repeated Addition

Row, Row, Row Your Line...

Grade Level: Middle

MATERIALS: No special materials are necessary

ORGANIZATION: Whole class activity

PROCEDURE: This is an ongoing activity which can be used as students are learning multiplication facts.

Typically, a classroom full of children do some type of lining up when entering or leaving the room for some activity. Instead of traditional lines, have students line up in arrays. For example, if lining up after lunch, announce to the class that you want them to line up in "3's," meaning 3 students across and as deep the number of students in the class. Before entering the room, walk down the columns and stand next to a particular row and ask a student to say the multiplication fact and product for where you are standing.

For example, if you are standing at the fifth row of students, the child would respond, "3 x 5 is 15." Move to several different locations and ask other students to respond.

Each time you line up, pose a different row structure for the students.

7	Represents Multiplication in Arrays and as Repeated Addition

Picture This

Grade Level: Middle

MATERIALS: Crayons, poster paper, meter sticks, glue, small slips of blank paper

ORGANIZATION: Individually

PROCEDURE: In this activity, students will complete a grid or poster which represents products in the form of repeated addition by drawing pictures of items which generally are found in sets of 1, 2, 3, etc.

In preparation for the lesson, brainstorm with students different items which usually come in sets. For example, sets of one might be noses or heads, sets of two might be eyes or scissors, etc. A list of possible sets is shown below.

Have students take a sheet of poster paper and write 1 through 9 across the top and 1 through 9 along the left edge. You may wish to have students draw lines to create a grid (this activity in itself can be used as a measurement lesson), or you may choose to have students try to glue pictures in straight lines which will look more like a poster.

Have students decide the sets of items they will use for their posters. We have found this lesson to be most successful if students are given small squares of paper on which to draw pictures rather than drawing them directly on the poster board. These small squares can then be pasted in place on the board.

1 - nose
2 - eyes
3 - traffic lights
4 - animal legs
5 - fingers on a hand
6 - dice
7 - days in a week
8 - octopus
9 -tic-tac-toe squares

Give students time to draw pictures and complete the poster. Usually, students will need two or three days to complete this project. Display students' posters when they are finished and have them refer to their art work when solving multiplication facts.

	1	2	3	4	5	6	7	8	9
1									
2									
3									
4									
5									
6									
7									
8									
9									

8	Uses Various Methods to Estimate Products

Baa, Baa, Black Bean
Grade Level: Middle

MATERIALS: Ziplock plastic bags, white beans, black beans

ORGANIZATION: Whole class activity

PROCEDURE: This activity involves estimating products. The teacher should explain to the students that traditionally, a shepherd would have one black sheep for every ten or for every one hundred sheep in his flock. In this way, the shepherd could more easily count the sheep

In this activity, students are going to toss a bean bag and use the method described above to estimate the number of beans per bag. To prepare for the activity, the teacher needs to fill bags with varying amounts of white beans in a set ratio to black beans to indicate amounts in the bag. A large Ziplock bag allows the beans to be spread out so that the black beans are obvious. The bag may be stapled or taped to insure it stays closed.

Have the students sit in a large circle and distribute a bag of beans to each student. On a signal from the teacher all students should toss their bags into the center of the circle. The students must line the bags up in numerical order from fewest to most beans. This activity may be timed and repeated many times to see if they can estimate and line up the bean bags faster than before.

The teacher may vary the amount that the black bean represents.

Hands On, Inc
2121 Rebild Drive
Solvang, CA 93463

8	Uses Various Methods to Estimate Products

Guessing Can Be a Factor!

Grade Level: Middle/Upper

MATERIALS: An index card for each student, a paper bag to use as a grab bag

ORGANIZATION: Whole class activity

PROCEDURE: This activity provides a fun approach to estimating products. Once the grab bag is prepared, it makes a good end of class activity to fill in a two or three minute block of time before the bell rings.

Give an index card to each student. They should cut or neatly tear the card in half. On one card they should write a one digit number, on the second card they should write a two or three digit number. All of the cards should be placed in a paper bag.

The activity is just a matter of allowing a student to reach in the bag and pull one card and announce the number. The class should be given a moment or two to think about how they might round off or change the number to help them quickly be able to estimate a product once a second number is given. For example, if the number is 89, they would round up to 90.

Select a group of four or five students sitting in a cluster and tell them that they will be the "solvers." Allowing the whole class to participate at one time can become chaotic As soon as a second card is pulled and announced, the solvers should each try to quickly estimate the product and announce the answer for the class.

Another student pulls a second card from the bag and reads it aloud. The solvers would respond with their estimates as explained above. The student who is closest to the actual answer should explain the method used for the estimation.

Hands On, Inc
2121 Rebild Drive
Solvang, CA 93463

8	Uses Various Methods to Estimate Products

On a Roll for the Tenth Time

Grade Level: Upper

MATERIALS: Dice (with stickers to be placed over the numbers) and calculators

ORGANIZATION: Groups of two

PROCEDURE: This activity gives students practice in mentally multiplying by multiples of 10.

Divide the class into groups of two and give each student a die and some small stickers. Have them "re-label" the die with the numbers: .001, .01, .1, 10, 100, and 1000. One student should write a number on a sheet of paper, the other student then rolls the die and must say the product of the written number times the digit (factor) on the die.

Students should take turns multiplying. If the student who wrote the number feels that the product given by the teammate is incorrect, the answer should be checked on the calculator. The game can be played by giving points for correct answers or can be done as non-competitive practice.

8	Uses Various Methods to Estimate Products

At Home On the Range

Grade Level: Upper

MATERIALS: Calculators, paper and pencil

ORGANIZATION: Teams of two students

PROCEDURE: This activity gives students practice at estimating multiplication of decimals in a gamelike setting.

Each group of two students should meet together and make a "game board" of decimal and whole numbers with four sets of adjacent ranges. Each range should have a span of at least two tenths. A sample board might look like:

Range 1	Range 2	Range 3	Range 4
.01 through .31	.32 through .6	.61 through .855	.856 through 1.5

Match one team against another. The goal of this activity is for the first team to write a 1 or 2 digit number for the second team (these may be whole or decimal numbers) and point to a range section. Team two then has two chances to estimate a factor which, when multiplied by the number given by team one, fits within the chosen range. Estimates should be given without using the calculator; but the calculator may be used after the first estimate to help with the second and final estimate. Teams should take turns writing numbers for one another.

As an example, imagine that team one writes 45 and points to range 2. Team 2 might estimate .0015 as the second factor. On the calculator, they would check to find that the product is .0675 -- too great. The second guess might be .0011. They would check on the calculator and compute a product of .0495 -- within range 2. They would receive a point for their correct estimate. Teams receive 2 points if their first estimate is correct.

9	**Performs Multiplication Computation Including Those that Require Regrouping**

Making a Square Deal

Grade Level: Middle/Upper

MATERIALS: Dice for each group, cards with multiplication frameworks (shown below)

ORGANIZATION: Cooperative groups of three or four

PROCEDURE: In this activity students have to do planning and analyzing of regrouping in multiplication to successfully fill in their multiplication chart.

Divide the class into groups and give each group a pair of dice. Next, have each student make a framework as shown in sample A below. Have the class discuss some of the limitations in the range of possible products. For example, since 6 is the largest number on a die, the greatest number that could possibly occur would be 4,356 (66 x 66). Since there are no zeros, the lowest product would be 121 (11 x 11). This discussion should take place prior to each new sample introduced. Once students understand, select one product for all students to use in game one.

To do the activity, the players take turns rolling the dice. On each roll, the player can select either of the digits rolled to write in one of the blank squares. Once a student writes a number in a square, it may not be changed

36	43	35
x 12	x 15	x 13
432	645	455

The next player rolls and all players repeat the process. For sample A, the dice would be rolled four times (the students select only one of the two numbers rolled each time). The following examples show three possible scenarios for dice rolls of 3 & 4, 3 & 6, 1 & 3, and 2 & 5. Imagine that the given product is 466. Of these three answers, 455 is the closest. The student who is closest gets to name the product for the group for the next round.

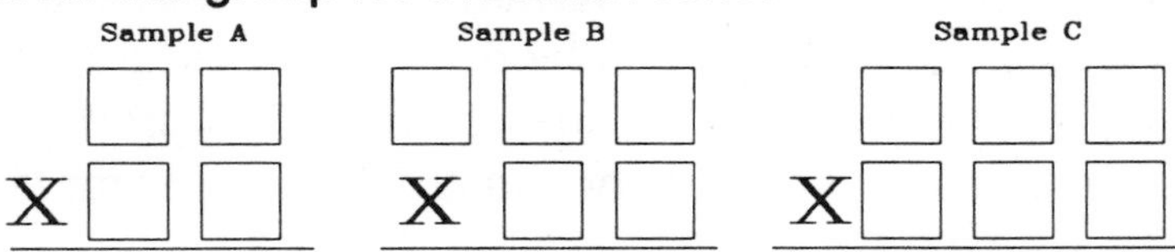

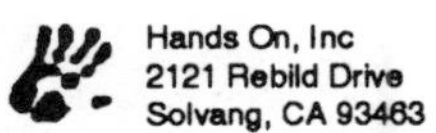
Hands On, Inc
2121 Rebild Drive
Solvang, CA 93463

9	Performs Multiplication Computation Including Those that Require Regrouping

Cut Time

Grade Level: Upper

MATERIALS: Graph paper and scissors

ORGANIZATION: Individually or in teams of two

PROCEDURE: Begin the lesson by directing students to cut out nine 10 by 10 grids, nine 1 by 10 grids, and nine 1 by 1 pieces from a sheet of graph paper.

Write a sample problem on the chalkboard. We'll use the example shown on the right. Have the students use their graph paper pices to set up this problem in a visual format. Note the zero (s) placed above the tens and hundreds columns.

348
x 26

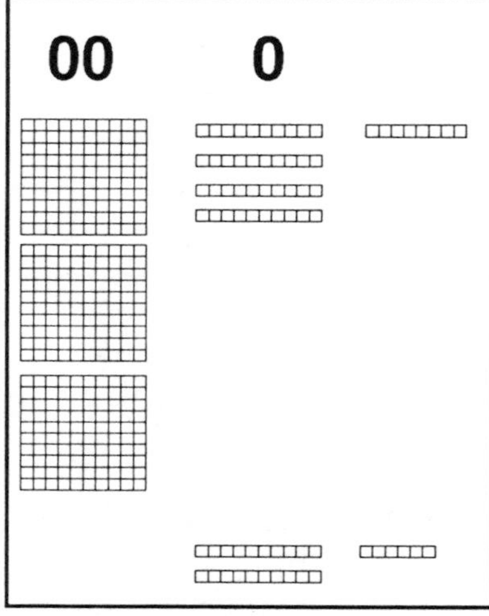

Ask students to look at the numbers in the ones column (8 and 6). "What is the product of 8 x 6?" Write this answer on scratch paper.

Ask, "In the number 348, how many tens are in the tens column?" (4) . "What is '4 tens times 6?'" (6 times 4 is 24 and since one number is a ten it is 240). Write this number.

Continue this questioning to include: 6 times 3 hundreds; 2 tens times 8; 2 tens times 4 tens; and 2 tens times 3 hundreds. Each time have students write their answers in a column.

48
240
1800
160
800
6000
9048

Ultimately, have them add the six numbers. When students understand how this process works, have them make up a multiplication problem. Using their graph paper cutouts, have them write out a step by step explanation of how to solve their problem using this method. These explanation can be pasted to a sheet of construction paper for classroom display.

Hands On, Inc
2121 Rebild Drive
Solvang, CA 93463

10	Identifies and Uses the Basic Elements of Division including Basic Facts, Terminology and Remainders

Beans Can Pay Dividends

Grade Level: Middle

MATERIALS: Beans, two dice for each team, paper and pencil

ORGANIZATION: Teams of two students

PROCEDURE: This activity provides a transition for students from concrete examples of basic division facts to the division algorithm.

Divide the class into teams of two and give each team two dice and thirty to forty beans. The students should draw one division algorithm diagram ($\overline{)\quad}$) which they will share.

Begin by having one student roll the dice. Using the two digits rolled, the student must then decide how the digits can be placed in the diagram to make a division problem which has no remainder. The student must explain the position of the numbers by using the correct terminology.

For example, if a 2 and 5 were rolled, the first student would say, "The divisor is 2, and the quotient is 5. The second student then places the corresponding number of beans in these two locations and must then place the correct number of beans in the remaining location (10 beans in the dividend). If the first student is correct, he/she receives a point, as does the second student if placement of beans in the third location is correct.

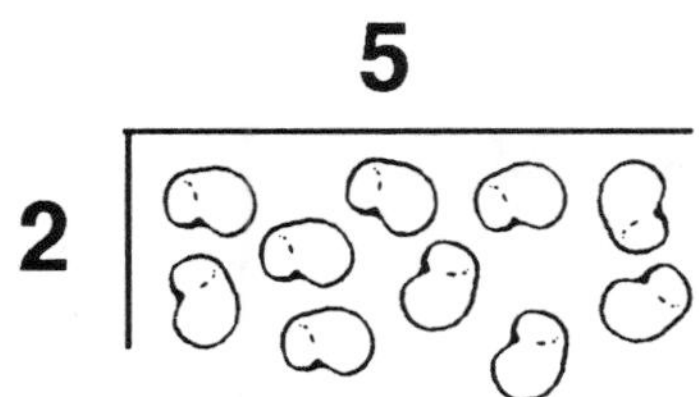

Should the first student use the incorrect terminology (for example if the student were to say, "5 is the divisor and 2 is the dividend," the problem cannot be done using basic division whole number facts). In this case, the second student could recite the correct placement and solve for the unknown to receive two points.

Students should take turns rollling the dice.

10	Identifies and Uses the Basic Elements of Division Including Basic Facts, Terminology and Remainders

Taking Division Cereally

Grade Level: Middle/Upper

MATERIALS: Ten different types of cereals (different colors, shapes, etc.), 10's number charts (Appendix D)

ORGANIZATION: Individually

PROCEDURE: This activity gives students an opportunity to find and describe patterns in divisibility of numbers between 1 and 100.

Duplicate copies of a 100 numbers chart (Appendix D) and distribute them to the students. Working with the whole class, display different cereal types. For example, some cereals have different colored pieces, some have different shaped pieces, etc. You should be able to find ten different shapes and colors in three different cereal boxes.

Have students make a "legend" of cereal types. Green spheres might represent numbers which can be divided by 2; blue stars might represent numbers which can be divided by 3, etc. Each student should use the same legend.

Have students place the cereal types on the squares which are divisible by that number. Once they have finished, have students volunteer to describe patterns that they find in divisibility. A sample response might include, "All the numbers in the five and ten columns have Cheerios, which means that all numbers that end in five or 0 are divisible by 5."

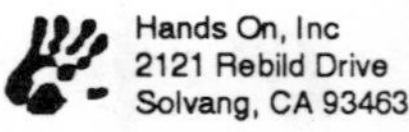

10	Identifies and Uses the Basic Elements of Division Including Basic Facts, Terminology and Remainders

Spin to Win

Grade Level: Middle

MATERIALS: Spinner for each team (Appendix C), paper and pencil, classroom clock with a second hand

ORGANIZATION: Teams of two students

PROCEDURE: This lesson provides practice in divisibility and compatible numbers in a game-like format.

Begin by dividing the class into teams of two. Have each team make a spinner as shown in appendix C. The spinner should be divided into ten sections and labeled with numbers 0 - 9. Each student will need a pencil and paper as well.

To play the game, player 1 spins the spinner four times, each time writing down the specified number. As an example, we'll imagine that a student spun 0, 7, 6, and 2. The student then has 30 seconds to create a division problem which uses as many of the numbers as possible WITHOUT leaving a remainder. Student may not use 0 as a divisor.

Player 1 receives 1 point for each number which is used in the problem. If no problem can be made, then no points are scored.

In our sample problem, student 1 might do any of the following:

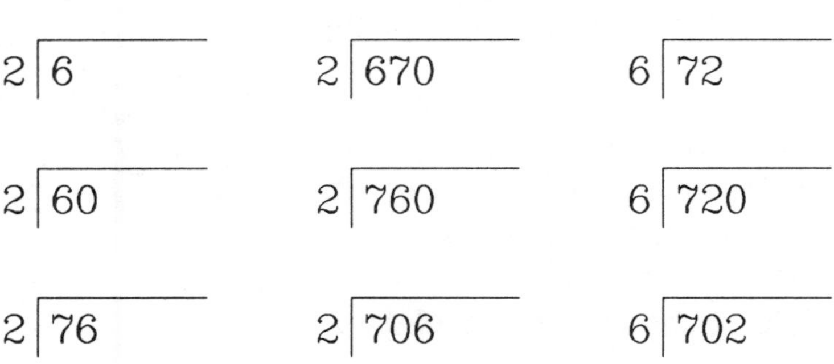

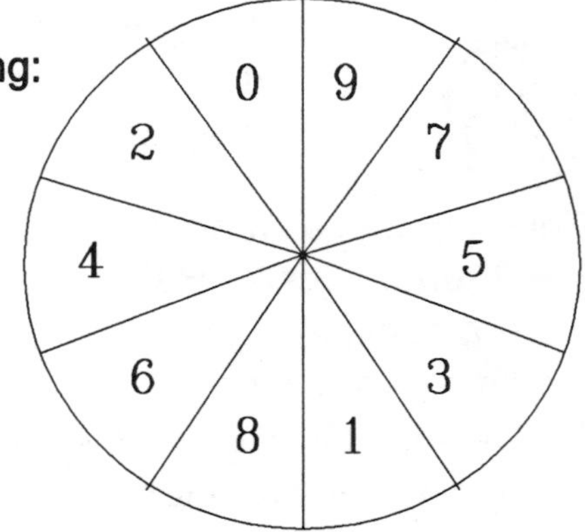

10	**Identifies and Uses the Basic Elements of Division Including Basic Facts, Terminology and Remainders**

Food for Thought

Grade Level: Middle/Upper

MATERIALS: Macaroni (four types), poster paper, glue

ORGANIZATION: Groups of four

PROCEDURE: In this activity students will be making posters to represent a division problem with all components (divisor, dividend, quotient, remainder) identified.

Begin by explaining the different components in a division problem. The "or" ending on **divisor** identifies it as the number that will do the dividing. Similarly, the ending "-end" is equivalent to the "-ent" ending of recipient which indicates it receives the action of the word, therefore; the **divid-end** is the number which will be divided by the divisor. The word **quotient** is from the Latin "quotiens" meaning "how many times" and shares the same root as quota which also refers to an amount. Quotient refers to how many times the divisor will divide into the dividend. The word remainder is simply what is left over; that part of the dividend that the divisor will not divide into evenly.

After students are familiar with these terms, distribute the macaroni to the students and have them create three division problems on poster paper. They should pick one type of macaroni to represent each of the four terms of division (divisor, dividend, quotient, and remainder); this information should be written as a "legend" or "key" at the bottom of their poster.

For example, a student may decide to represent the algorithm 12 divided by 5 = 2 remainder 2. He or she might glue twelve small elbow macaroni and write a division sign to the right (or division bar below). Next, the student would glue five straight noodles to the right of the division sign, two seashell noodles for the quotient and two rotelli noodles for the remainder. Use these as displays in your classroom to reinforce the vocabulary of division.

11	Identifies Addition/Subtraction and Multiplication/ Division as Inverse Operations

Handy With the Beans

Grade Level: Middle/Upper

MATERIALS: Beans or other counters

ORGANIZATION: Groups of two

PROCEDURE: This activity provides subtraction and addition practice of basic facts through 18.

The teacher should demonstrate the procedure to the class by placing 18 beans in a pile and taking a number of beans into the right and left hands. Some beans should remain in the pile.

A class member should count the total number of beans remaining in the pile and subtract this number from 18 (thus discovering the total number of beans in the right and left hands). The teacher then displays the number of beans in the right or left hand and students must then quickly compute the number of beans which remain in the opposite hand.

When students understand the activity, have them work in pairs and take turns taking handfuls of beans from a pile of 18.

Several variations of this activity are possible, including students writing equations to represent the missing numbers or students using various numbers of beans in the original pile and discovering the amount by subtracting the number of beans in the right and left hands.

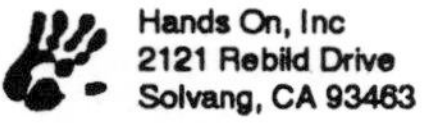
Hands On, Inc
2121 Rebild Drive
Solvang, CA 93463

11	Identifies Addition/Subtraction and Multiplication/ Division as Inverse Operations

Opposites Are Attractive

Grade Level: Middle/Upper

MATERIALS: Clay or bread (something to cut), popsicle sticks or plastic knives (to cut with)

ORGANIZATION: Teams of two

PROCEDURE: In this activity students will physically demonstrate the concept that certain operations are opposites. In addition they will learn that opposite operations can be used to check one's work.

Hand out the sticks and clay in strips about ten inches long so that each group has enough for experimentation. Explain that frequently people perform calculations without bothering to check their work to see if it's correct. Today students are going to find a simple way to check their answers.

Have them begin by setting up their desk so that their partner can't see their work. Write a number on the board for everyone to use. As an example, we will use 12. Have them make twelve marks in their strip of clay fairly evenly apart. Next tell them to subtract, or cut, an amount of their own choice from their strip. After they have done this, they should give their partner one of the separated sections. The job is for each partner to figure out how much of the original strip their partner still has. Ask them to figure out how much they would have to ADD to make the piece they have been given into a strip of twelve.

Point out that they are using addition to get back to original number—12. Students should realize that addition is used to check subtraction and vice versa. Let them experiment with this process until they are comfortable.

This same approach can be used for multiplication and division. You may want to put a larger number on the board for students to use.

Hands On, Inc
2121 Rebild Drive
Solvang, CA 93463

11	Identifies Addition/Subtraction and Multiplication/ Division as Inverse Operations

Operation Collage

Grade Level: Middle/Upper

MATERIALS: Various magazines and newspapers, glue, scissors, poster/construction paper

ORGANIZATION: Individually

PROCEDURE: It is important that students recognize that certain operations are the opposite/inverse of others; it will help them mentally check calculations and estimations. In this activity, students will make collages to represent opposite operations.

Review with your class which operations are opposites and how to check work using that information. Now explain that each student is going to create a simple collage which will represent some mathematical operation. The class will then analyze each of the collages and write down (or create another collage) the opposite, but equal, operation.

For example, two pictures of motorcycles might equal a car (2 tires on the motorcycle x 2 = 4 tires on the car). The opposite of this would be 4 divided by 2 = 2; the same numbers are used in an equal but opposite operation. This could also be represented by a collage depicting a cow cut in half (divided by two) equaling a bird, human, or other two-legged animal. It may be difficult to find pictures which link together completely logically; therefore, you may wish to simply focus on the quantities involved.

When the students have finished their collages, display them in the room and have each student write down the opposite operation using the same numbers to check the work. Discuss the results and elicit generalizations from the students.

11 Explains Addition/Subtraction and Multiplication/ Division as Inverse Operations

Explaining Inverse in Straws

Grade Level: Upper

MATERIALS: Straws, rulers (centimeter measures), scissors, glue, and construction paper

ORGANIZATION: Groups of two or three

PROCEDURE: In this activity, students will make physical models to represent inverse operations.

Begin the lesson by discussing the meaning of the words "inverse" and "operation." Students should simply be able to identify that an operation is adding, subtracting, multiplying, or dividing and inverse means opposite. As an example, write: 6 + 3 = 9 and 6 x 3 = 18 on the chalkboard. Select a student to explain how these equations could be rewritten, using the same numbers, to represent a subtraction and division equation, respectively. Help students with their explanations.

Divide the class into groups and give each group several straws and scissors. Tell them that they need to use straw sections to represent either the equations given above or a set of their own. To accomplish this, they should place the straws next to a ruler or centimeter rule and make marks at equidistant lengths. Let them figure out the scale that they want to use.

Next, have them cut lengths of straw to represent the numbers in the equations and glue them on a sheet of construction paper. They should also write an explanation under the addition/ subtraction model and the multiplication/division model. Have groups of students display and explain their posters to the class.

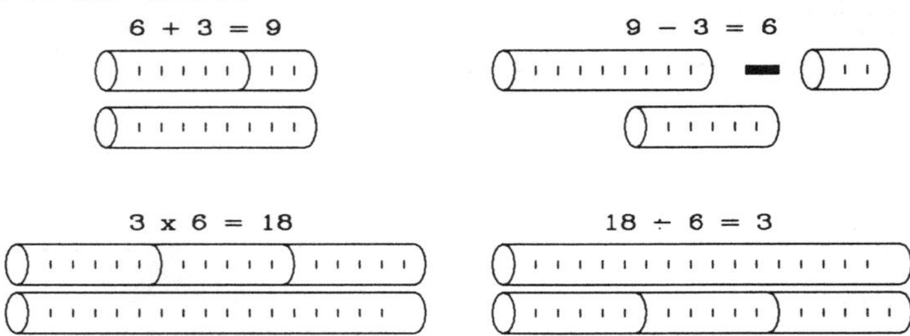

12 | Uses Various Methods to Estimate Quotients

A Parting Guess

Grade Level: Middle

MATERIALS: No special materials are necessary

ORGANIZATION: Whole class activity

PROCEDURE: This lesson takes estimation of quotients to a practical level in which students will be estimating how to divide lengths and distances in equal parts. Rather than a one day lesson, it is ongoing and can be used for a minute here and a minute there.

Students usually think of division as a paper and pencil activity in which they follow the steps of the algorithm -- divide, multiply, subtract, and bring down. Point out to students that they use division in many other ways. For example, whenever they say, "I'm almost halfway." or "about a quarter of the class," they are using estimation in division.

There are numerous activities which the teacher can do throughout the day to have them practice this skill of estimation. For example, have them draw lines to divide their paper into equal sections (use a different number each time); have them estimate how to divide the playground into x equal grids; have them divide their desk into equal groups; the possibilities are unending.

This type of estimation allows students to physically and visually see what happens when things are divided and to establish a sense of ratios in their minds.

12	Uses Various Methods to Estimate Quotients

Groupies

Grade Level: Middle

MATERIALS: Small items in bags which students bring from home (marbles, buttons, paper clips, etc.) which can be divided into groups

ORGANIZATION: Whole class activity

PROCEDURE: This is a beginning lesson for teaching division in which each child has time to prepare a presentation of how the items in a bag can be divided.

On the day before the presentations, assign the homework of having each child bring a baggie with 15 to 25 items. The items should be the same to avoid confusion with equal groups.

On lesson day, begin by demonstrating the process for the children. Display a bag of items and show how these can be separated into groups of various sizes. The easiest way to demonstrate this is to "deal" the items into groups of three, four, five, etc. Show students that in many occasions, the items do not divide "evenly." Tell them that it is their task to find group numbers for their items into which they do divide evenly.

When students understand the concept, have them work individually to repeat this process. Encourage students to use groups larger than two or three. Have them record the various sizes of groups that divide evenly.

Students should prepare a mini-presentation for the class in which they share their items and then explain the types of group configurations they discovered. Have children estimate the number of groups that will divide evenly or the number of items that would be in each group prior to each student's explanation.

Hands On, Inc
2121 Rebild Drive
Solvang, CA 93463

12	Uses Various Methods to Estimate Quotients

Don't Let Squares Divide Us!

Grade Level: Middle/Upper

MATERIALS: Two dice for each group, paper and pencils

ORGANIZATION: Cooperative groups of four

PROCEDURE: This activity gives students an opportunity to do estimation and place random numbers to achieve a desired goal.

Divide the class into cooperative groups of four and give each group a pair of dice. Students should begin by making a chart as shown in Sample A. Before beginning the activity itself, discuss with the students the range of quotients which might occur using only the numbers 1 - 6 as divisor and dividend. Generally speaking, any digit could occur in the quotient; however, since there will be no 7's, 8's, 9's or 0's in the divisor or dividend, the quotients suggested by the students are best kept to numbers using the digits 1 - 6.

To begin the activity, a student in each group should predict a possible 1 or 2 digit quotient which each member of the group writes in the appropriate places in their blank algorithm. The first student rolls the dice and each student selects one of the digits rolled to place in one of the empty digit squares. Once a number has been written in a square, it cannot be changed.

After four rolls, the squares will be filled. At this point, the student who has most closely approximated the given quotient is the winner.

As students become more sophisticated with the activity, you will find their ability to estimate will improve as will their understanding of the step by step process of doing a division problem.

Sample A Sample B

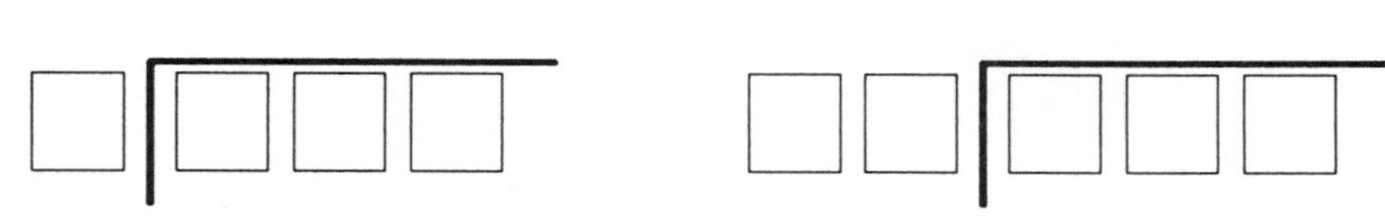

Hands On, Inc
2121 Rebild Drive
Solvang, CA 93463

12	Uses Various Methods to Estimate Quotients

Column As You See 'em

Grade Level: Upper

MATERIALS: Calculators for two student judges

ORGANIZATION: Divide the class into two equal teams

PROCEDURE: This activity provides a trial and error approach to practice estimating quotients.

The teacher should begin by listing two columns of numbers on the chalkboard. A sample list might be: column I: 2, 4, 6, 8, 3, 5, 8, 10, 11, 12, and 15; column II: 36, 64, 35, 43, 78, 56, 87, 98, 100.

Select student judges to be timers and check accuracy. Ask for a volunteer from team 1 and tell the students that the volunteer will have five seconds to estimate a quotient of a number in column II divided by a number in column I. The teacher will point to the two numbers. If the student is correct, the team gets 10 points. One point is deducted from 10 for each digit above or below the actual answer.

For example, if the teacher were to point to 4 and 36 and the student responded "9," the answer is exact so the team would score 10 points. If the student guessed "7," then the team would receive 8 points (7 is 2 less than 9 and 10 - 2 = 8).

The teacher would then point to a volunteer from team 2 and would point to two different numbers. Continue until each student has had a chance to respond.

As an extension, you can use larger numbers or use numbers containing decimals.

Hands On, Inc
2121 Rebild Drive
Solvang, CA 93463

13	Performs Division Computations

A Dime is Two Nickels Is Ten Pennies

Grade Level: Middle/Upper

MATERIALS: Play money coins

ORGANIZATION: Individually or in teams of two

PROCEDURE: Since most students are familiar with money and coins, this activity provides division practice along with teaching equivalent money values.

Begin the lesson by demonstrating the activity for the entire class. You may find that you will need to explain two or three different situations before the students fully understand their task.

Write the value $.40 on the chalkboard and ask students to decide on the various ways in which coins totaling 40 cents could be made into equal groups. Elicit various responses including: 4 dimes, 8 nickels, 40 pennies, 2 groups of 2 dimes, 2 groups of 4 nickels, four groups of 2 nickles, 8 groups of 5 pennies, etc.

Record all responses on the chalkboard and then have students come forward to write the division equation and algorithm which would depict each grouping ($40 \div 10 = 4$).

When students understand, have them divide into teams of two. Each group should select a different total amount (they should choose an amount that can be divided evenly) and then make a chart which depicts the various combinations which would equal their total. They should display it in poster form (shown below).

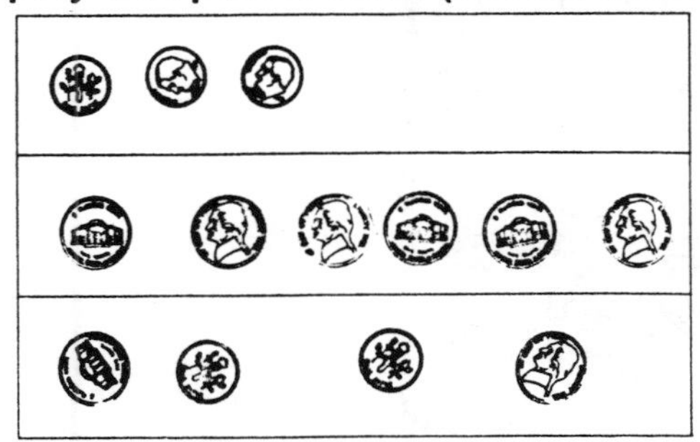

At this point, have the students become the teacher. They should display their charts and one team at a time call on classmates to come forward and write equations and the division algorithm for various grouping described on the poster.

Hands On, Inc
2121 Rebild Drive
Solvang, CA 93463

13	Performs Division Computations

Lunch Bunch Munch
Grade Level: Middle/Upper

MATERIALS: Grocery store flyers or advertisements, glue, scissors, art paper

ORGANIZATION: Groups of four or five

PROCEDURE: This is an activity for practice in addition and division using the basic food groups.

Begin by discussing basic food groups with the whole class. A poster or health/science book is helpful.

The students will organize a lunch for the group. They will do this by cutting out food pictures from the ads and pasting them on the art paper, making sure the four food groups are included.

They should record the cost of each item used and compute the cost per person.

You might have students use calculators to do their computation.

13	Performs Division Computations

Calculating the Great Divide

Grade Level: Middle/Upper

MATERIALS: Calculators for each student

ORGANIZATION: Individually

PROCEDURE: Students are often familiar with the concept that multiplication is actually a process of repeated addition; fewer students are aware that division is a process of repeated subtraction. This activity uses calculators to demonstrate this point.

Distribute calculators to each student and write a simple division problem on the board. Tell students that rather than performing the division in the traditional manner on the calculator, you want them to divise a method for solving this problem using only the subtraction key. Discuss with the class how this might be possible.

Give the students time to experiment and then call on three of four students who were successful to explain how they got their answers and then to explain why this method worked.

Once students understand, have them write a paragraph explaining the relationship of multiplication and addition versus division and subtraction.

Hands On, Inc
2121 Rebild Drive
Solvang, CA 93463

13	Performs Division Computations

A Place for Everything
Grade Level: Middle/Upper

MATERIALS: Small cards (index cards cut into quarters), envelopes, full-size index cards

ORGANIZATION: Individually

PROCEDURE: In this activity, each student will make a set of cards which depict a division algorithm. The cards will then be rotated around the room with each student trying to place the cards in their correct order to solve the division problem.

Distribute a set of 12 cards to each student and on a separate sheet of paper, have the students create and solve a division problem with a one or two digit divisor (the teacher should decide which to use and each student should create the same type of problem; for example, a two digit divisor and a two digit dividend). Students should ask a partner to check that the computation is done correctly.

At this point, the student should write the problem with the solution on a full size index card and place this card in the envelope. The students should also write one of the numbers of the solution on each of the small cards. On the cards representing the numbers used for the dividend, the students should write "DIVIDEND."

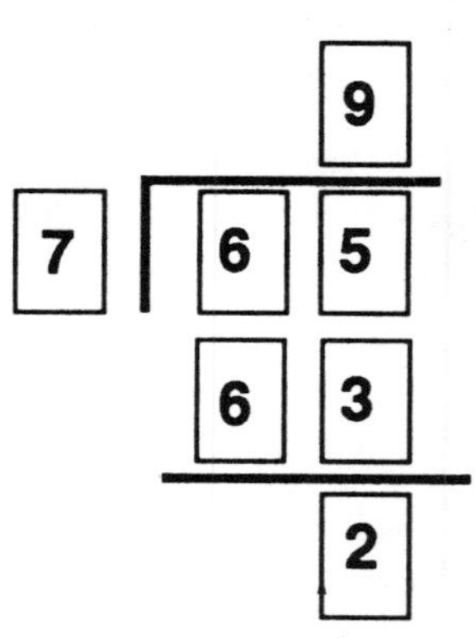

Place all of the necessary cards in the envelope and set up a rotation system around the classroom. Give students time to try to place all of the cards in their proper location and then check the index card to see if their solution is correct. Before rotating the cards to the next player, have the class discuss the strategies they used to solve the puzzle, such as looking for multiplication combinations, counting the number of cards to figure if there would be a one or two digit quotient. Continue passing envelopes with sets of cards as students learn more about how division works.

14	Identifies and Writes Addition, Subtraction, Multiplication, and Division Problems in Vertical or Horizontal Formats

A Dicey Operation

Grade Level: Middle

MATERIALS: Dice, small stickers, and beans

ORGANIZATION: Cooperative groups of two

PROCEDURE: This is a game for practicing all operations. Students will need two dice per group. One regular die, and one with stickers on the faces labeled +, +, -, -, x, and ÷ .

The teacher should demonstrate by playing one game with a volunteer and begins by placing a pile of approximately 100 beans in the center of the table. Each player should take ten beans. Roll the dice and then do the operation shown on the die -- adding, subtracting, multiplying, or dividing the number of beans they have by the number shown.

For example, if a 4 is rolled with a +, the teacher takes four beans from the center; if a - is rolled, four beans are placed back in the center; if a x is rolled, the teacher takes 40 beans (10 x 4); and if a ÷ is rolled, the student gives up 2 beans (10 ÷ 4). If the operation cannot be performed then the dice should be re-rolled.

The student now takes a turn rolling the dice. The game is won in one of two ways. A player rolls a number which allows him or her to take all of the center beans and then all of the other players' beans as well, OR a player rolls a - or ÷ and doesn't have enough beans to place back in the pile. The player with the most beans wins.

Once students understand, have them divide into pairs and play the game.

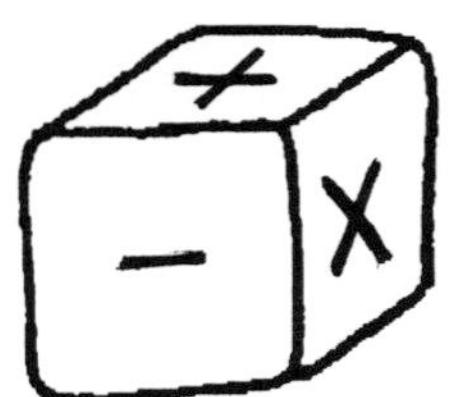

Hands On, Inc
2121 Rebild Drive
Solvang, CA 93463

14	Identifies and Writes Addition, Subtraction, Multiplication, and Division Problems in Vertical or Horizontal Formats

Match My Deck

Grade Level: Middle/Upper

MATERIALS: Index cards or some other type of card on which students can write (20 cards per student)

ORGANIZATION: Individually and then in teams of two

PROCEDURE: In this activity, students will create a set of 20 cards to be used in a "concentration" type game. Ten of the cards will have different problems written in different formats which will match the answers to the problems written on a second set of ten cards.

Begin by writing the examples shown to the right on the chalkboard and explain that there are several ways in which a problem may be written. Their task will be to make ten sets (pairs) of these cards. Each set should depict a method of displaying a math problem, each problem should have a different answer, and each problem must be written in a different format. Students may generate formats which are not shown below (including problems written in expanded notation, scientific notation, or other other creative formats).

$$\begin{array}{r} 24 \\ +16 \\ \hline \end{array} \qquad 37 + 13 =$$

Since students will be playing a concentration type game with these cards, the answers must be simple enough that they can be computed without paper and pencil.

24 + 36 =	60	3 x 12	36	$\frac{18}{6}$	3
$\begin{array}{r} 18 \\ +12 \\ \hline \end{array}$	30	$\begin{array}{r} 4 \\ \times 4 \\ \hline \end{array}$	16	$\begin{array}{r} 4 \\ 4 \\ 3 \\ \hline \end{array}$	11
16 − 6 =	10	45 ÷ 9 =	5		
$\begin{array}{r} 30 \\ -22 \\ \hline \end{array}$	8	$3\overline{)300}$	100		

Once students have finished their set of cards, have them choose a partner to play a game of concentration with a classmate (cards are placed face down in rows and columns and children take turns turning two cards in hopes of finding a match). A sample set of cards with matches is shown to the left.

Hands On, Inc
2121 Rebild Drive
Solvang, CA 93463

15	Understands and Uses the Terminology of Fractions

Just a Fraction...More or Less

Grade Level: Middle/Upper

MATERIALS: Dice, paper and pencil

ORGANIZATION: Groups of two

PROCEDURE: This activity challenges students to identify relationships between fractional values by deciding if proper fractions are greater or less than one-half and whether improper fractions are greater than or less than one and one-half.

Divide students into pairs and assign one member of each pair to record proper fractions, the other to record improper fractions. Students should label a paper with their given task and then draw a line down the center of the page, labeling the pages as shown below. The teacher may need to review the concept of proper and improper fractions, the relationship of the numerator to the denominator in each case, and strategies for determining whether a fraction is greater or less than one-half or greater than or less than one and one-half.

Provide a pair of dice for each team and have students take turns rolling the dice, each time making both a proper and improper fraction from the two digits. They should then write their fractions in the appropriate columns.

Have a student record ten to twenty fractions and then have them write a statement which explains how they can tell if fractions are greater than or less than one-half or one and one-half.

Proper	
> 1/2	< 1/2

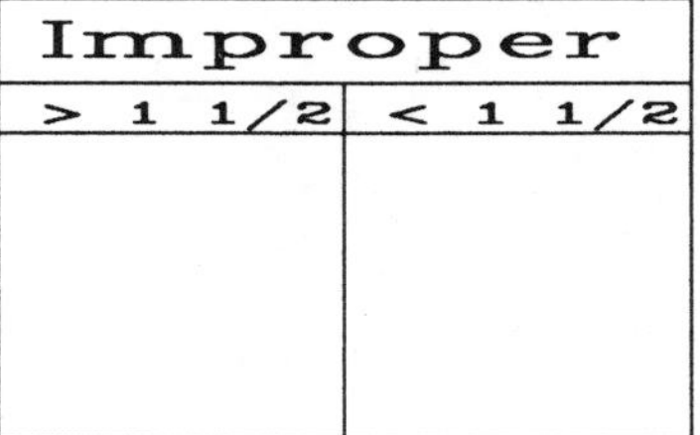

Improper	
> 1 1/2	< 1 1/2

15	Understands and Uses the Terminology of Fractions

Find the Factors, Old Bean!

Grade Level: Upper

MATERIALS: 30 to 50 beans for each student, paper and pencil

ORGANIZATION: Teams of two

PROCEDURE: In this activity, students will grab a handful of beans from a container and then find multiples of the number of beans they grabbed. It will give them practice in working with both factors and multiples

Place a large number of beans in five to eight large containers (placed at tables around the room). Ask for a volunteer to model the procedure. Have the student reach into one of the containers and grasp a handful of beans. Lay the beans out on a table or overhead projector and count them. Let's imagine that the student selected 38 beans.

Ask the class if they can think of any numbers that will divide evenly into 38 and record their responses. The full list of factors would be 1, 2, 19, and 38. Tell students that they will receive one point for each factor they correctly identify BUT if they choose a factor which is incorrect, they receive zero points for that turn. Should they choose a number of beans that is a prime number, and correctly identify it, they receive an additional five points.

Divide the class into teams of two and let them begin playing the game.

15	Understands and Uses the Terminology of Fractions

Placing an Order

Grade Level: Upper

MATERIALS: One inch graph paper and envelopes

ORGANIZATION: Cooperative groups of four

PROCEDURE: This activity promotes an understanding of improper fractions.

The teacher needs to prepare a set of four envelopes for each group. In each envelope, the teacher should place enough one inch squares to form a true square plus one or more extra squares (see below), The teacher should also make four cards for each envelope labeled with improper fractions that matches the contents of each envelope.

Review the concept of placing smaller squares together to form a larger square with the students to be certain they understand what they will be doing.

Distribute four envelopes and matching cards to each group. Each student should take one envelope and empty its contents on the table. Using the squares, they should make as large a square as possible (some will be left over). Ask students to discuss how they could determine the denominator (the number of small squares which is the size of the larger square) and numerator (total number of squares) of their improper fraction.

Once each student determines these numbers, have them select the appropriate card from the set of four. Ask students to explain how their improper fraction might be renamed to be written as mixed numbers. Exchange sets of envelopes among groups and repeat the process.

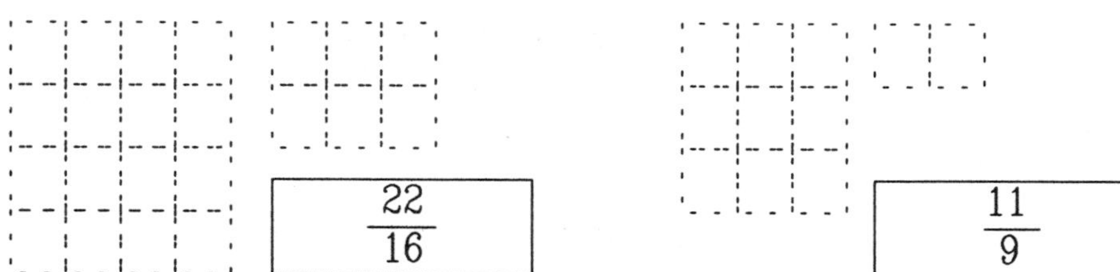

15	Understands and Uses the Terminology of Fractions

Somewhere, in There, Is a GCF!

Grade Level: Upper

MATERIALS: Four dice per group

ORGANIZATION: Students work in pairs

PROCEDURE: This activity gives students practice in quickly identifying the greatest common factor and can also be used to find the lowest common multiple as well.

Before beginning the activity, you may need to review the concept of GCF with the students. Divide the class into teams of two and give each team four dice. Students will take turns rolling the dice and arranging the four digits into two 2-digit numbers which have a greatest common factor.

For example, if a student were to roll 1, 3, 4, and 6; the digits could be arranged as follows:

13 and 46 or 64	31 and 46 or 64
14 and 36 or 63	41 and 36 or 63
16 and 34 or 43	61 and 34 or 43

Ask students to decide which of these configurations have a GCF. Students need only make one set of two digit numbers and then identify the GCF. However, students receive points for the largest GCF they can find. In this example, the student might identify 14 and 36 with a GCF of 2. In this case, the student would receive 2 points.

Students will learn to use a strategy in which they look for a larger (and more difficult) GCF in order to score more points. If the student incorrectly identifies a GCF or is unable to find a combination with a GCF, the other player has an opportunity to successfully find a combination.

The player with the highest point total after five turns is the winner.

16	Understands Ordering and Rounding Off of Fractions

A Fraction of an Order

Grade Level: Middle

MATERIALS: Construction paper squares or circles, scissors, paste, construction paper, markers

ORGANIZATION: Individually or in groups of two

PROCEDURE: This activity gives students practice in ordering fractions with unlike denominators.

Give each students a supply of ten to twelve construction paper circles or squares. Begin with the simple task of having them cut one of the squares to represent the fraction 1/2. Next, do 1/4, guiding any of those students who are having difficulty.

Once students understand the concept, have them cut shapes to represent: 1/5, 1/3, 5/8, 3/5, 2/3, 3/4, 3/8, 5/6, and 7/8. They should label each shape as it is cut.

Next, have students paste the shapes onto construction paper in order of size. Using this "number line," have them write five observations about how fractions are ordered.

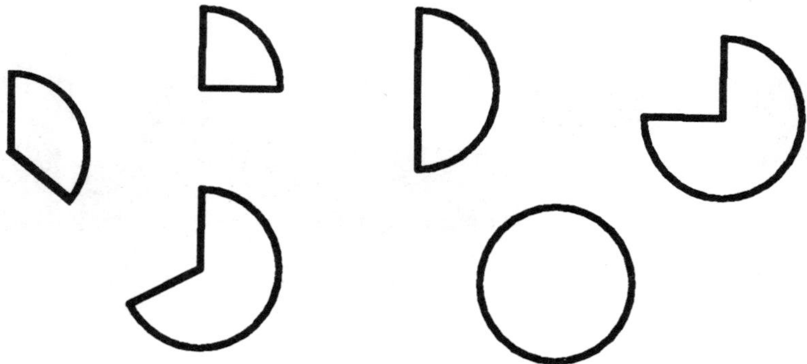

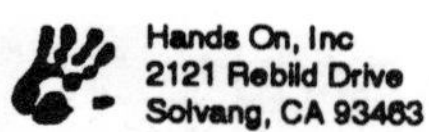

Hands On, Inc
2121 Rebild Drive
Solvang, CA 93463

16	Understands Ordering and Rounding Off of Fractions

Making Order of Fractions
Grade Level: Middle

MATERIALS: Construction paper, scissors, and paste

ORGANIZATION: Individually

PROCEDURE: This activity gives students practice in creating various fractions and then ordering them from least to greatest.

Have students suggest eight to ten fractions (avoid any mixed numbers or improper fractions) and write them on the chalkboard or overhead. Distribute construction paper, scissors, and paste and have students cut eight to ten congruent shapes of their choice (circle, square, rectangle, etc.) Using one of these shapes, have them trace ten outlines of their shape in a row (as shown).

Have them cut a representation for each of the fractions listed on the chalkboard and then arrange the fractions in order from least to greatest on a large sheet of construction paper. The fractional part should be pasted within each drawn shape. They should label the fraction represented by each cutout.

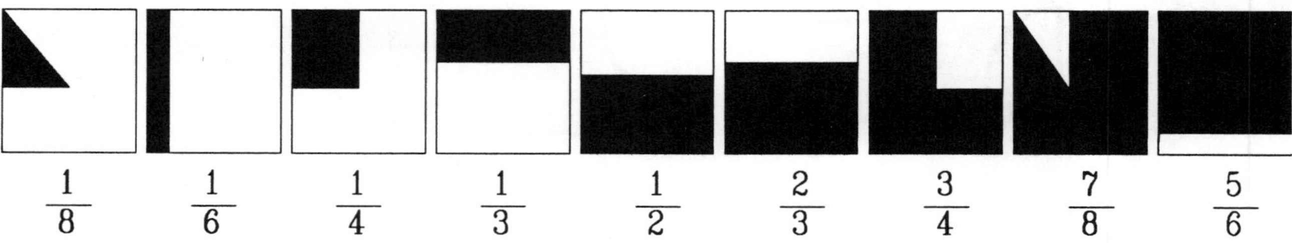

Hands On, Inc
2121 Rebild Drive
Solvang, CA 93463

16	**Understands Ordering and Rounding Off of Fractions**

A Round of Rounding

Grade Level: Middle/Upper

MATERIALS: Two dice for each pair of students

ORGANIZATION: Teams of two students

PROCEDURE: This activity gives students a game type situation in which to practice rounding fractional numbers.

Provide some background for this activity by writing: 0 1/2 1 on the chalkboard or overhead. Tell students to recall what they know about rounding numbers. In this activity, they will be rounding fractions to the numbers written on the board.

Roll two dice and tell the students the digits, for example, 6 and 2. With these digits, students should quickly make a proper fraction (2/6) and then tell whether 2/6 is closer to 0, 1/2 or 1. In this case, it is closer to 1/2. Students should then be able to tell how "far" (the difference between) the fraction and 0 and 1/2. A proper student response is: "2/6 is 2/6 larger than zero but it is only 1/6 less than 1/2."

Divide students into groups of two and have them roll the dice, taking turns to first say the fraction; next, identify the number it would round to; and finally, to tell the difference between its lower and upper range. If the student is correct at all three responses, a point is awarded. If the student is incorrect, the partner can win the point by answering correctly. If there is a discrepancy as to the right answer, students may challenge one another to draw a figure representing the fraction to prove that they are correct.

As an extension, have the students use four dice to practice working with larger digit fractions.

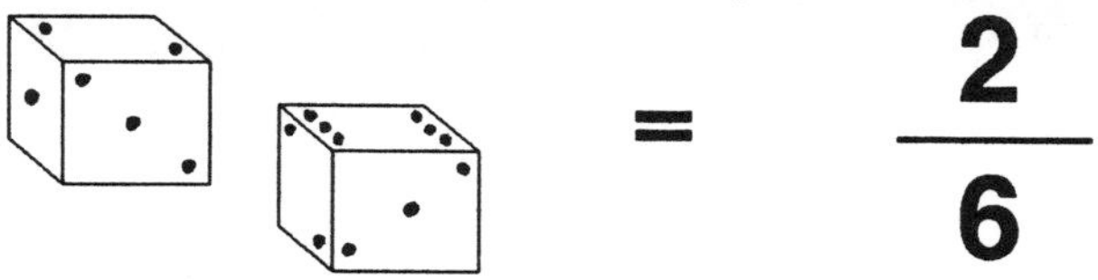

16	Understands Ordering and Rounding Off of Fractions

Draw, Pardner

Grade Level: Middle/Upper

MATERIALS: Construction paper and pencils

ORGANIZATION: Individually

PROCEDURE: One skill which will help students when working with fractions is to visualize the fraction with which they are working. This activity gives students practice in drawing and cutting fractional parts.

In most primary classes, students draw and cut 1/2, 1/4, and 1/3 of a shape. These fractions are fairly easy for older students to visualize. However, portions such as 3/7, 7/10, or 4/9 are less likely to be familiar to students.

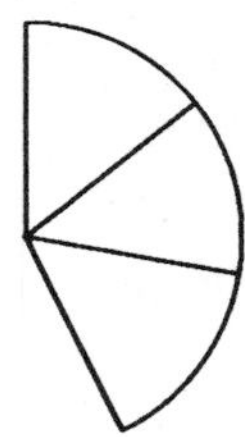

Distribute small pieces of construction paper to students. Rather than drawing a circle and then dividing it into fractional parts, have students draw only that portion of a circle described in a fraction suggested by a classmate. For example, if a student were to suggest 3/7, the class should draw the shape as shown to the right. Point out that 3 of 7 parts is a bit less than 1/2.

Continue this process, giving each student an opportunity to name a fraction to be drawn by classmates. You will probably be surprised at how difficult this activity will be for many of your students. We find that students need to practice this skill of drawing fractions many times before they can do it instinctively.

Although a circle is the easiest shape for students to draw, have the students experiment with drawing the fractional parts of squares, rectangles, triangles and other polygons.

Hands On, Inc
2121 Rebild Drive
Solvang, CA 93463

17	Reduces Fractions and Identifies Equivalent Fractions

Odd Man Out

Grade Level: Middle/Upper

MATERIALS: Decks of 49 cards (Appendix E)

ORGANIZATION: Groups of four

PROCEDURE: Duplicate and cut cards in Appendix E. There should be a set of 49 card for each group. Include 2 of each card in the appendix and add one more random card thus creating one card which will not have a match.

Begin by explaining that this game is similar to Old Maid in that the students will be trying to make "books" (sets of two matching cards) without getting stuck with the "Odd Man Out" card at the end of the game. The "books" or sets are made of equivalent fractions.

Divide the class into groups of four and deal all cards. Each child, in turn, takes a card from the player on the left. If a book can be made by the player, the two cards must be displayed for the entire group. If consensus is that the cards are indeed equivalent fractions, the book is placed on the table in front of that player. If it is not a match, the player keeps the cards in his/her "hand."

$\frac{1}{2}$	$\frac{5}{10}$

$\frac{3}{4}$	$\frac{6}{8}$

$\frac{1}{3}$	$\frac{3}{9}$

$\frac{1}{4}$	$\frac{3}{12}$

The game continues in this manner until one child is left with one card -- "Odd Man Out." Of the other three players, the student with the most books is the winner.

As an extension, this game can be played with equivalent percentages or decimals.

Hands On, Inc
2121 Rebild Drive
Solvang, CA 93463

17	**Reduces Fractions and Identifies Equivalent Fractions**

Cut it Out

Grade Level: Middle/Upper

MATERIALS: Construction paper, straight edge, paste, and scissors

ORGANIZATION: Individually or teams of two

PROCEDURE: In this activity students will be cutting various geometric shapes to represent equivalent fractions. These shapes will then be assembled to make a poster.

Begin the lesson by giving each student a square, a circle, and a triangle. Have students cut the square to represent 1/2; cut the circle to represent 2/4, and cut the rectangle to represent 3/6. Have them also cut out "=" symbols and paste these onto a large sheet of construction paper. Work through this portion of the lesson with the students.

Tell students that their challenge will be to complete a set of three more shapes to represent equivalent fractions for 1/3, 3/4, and 3/8 each. Students should, at a minimum, use a square, circle, and rectangle to represent a different equivalent for each fraction. Challenge students to try using more geometric shapes for more equivalent fractions for each set.

Have children paste their cut-outs on paper and display them as posters.

1 = 1 = 1

$\frac{1}{2}$ = $\frac{2}{4}$ = $\frac{3}{6}$

17	Reduces Fractions and Identifies Equivalent Fractions

If the Fraction Fits

Grade Level: Middle/Upper

MATERIALS: Scissors, crayons, fraction outline sheets (Appendix F)

ORGANIZATION: Whole class activity

PROCEDURE: This is an early lesson in finding equivalent fractions.

Distribute a copy of Appendix F to each students and have them shade each circle a different color. Each circle should then be cut out and fraction wedges should also be cut into sections.

Have student follow along as you give directions. First ask them to find two pieces with a common denominator (such as 1/4 and 1/4). Ask how many pieces they have (2). Have the students experiment with placing the two wedges they have chosen on top of other fraction wedges. They should try to find a wedge section which is equivalent to 1/4 and 1/4 (1/2).

Once students have done this, select several of them to come forward and write an addition sentence which describes the equivalents they've found (1/4 + 1/4 = 1/2). Use the problems generated by the students as models. Have each child prove whether or not the number statements written on the board are true.

Pose new and different combinations, each time letting students discover the relationships between the fractional wedges.

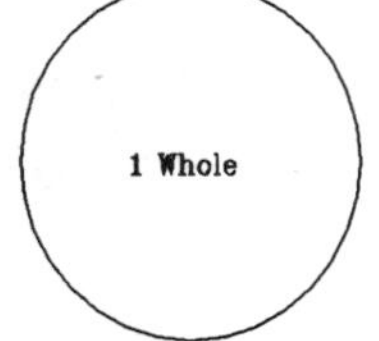

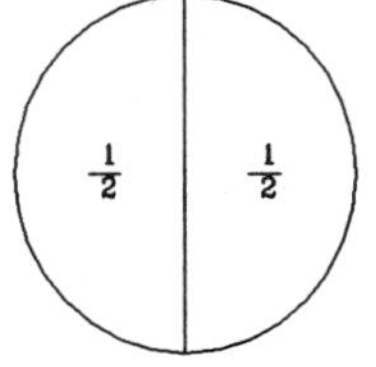

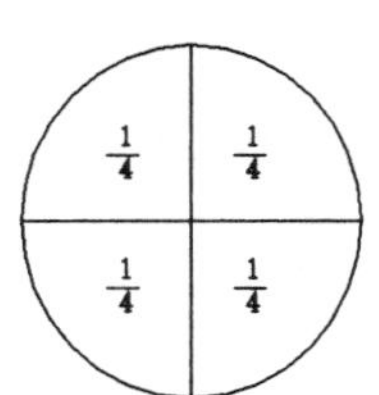

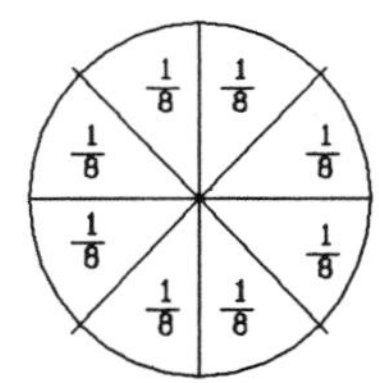

17	Reduces Fractions and Identifies Equivalent Fractions

Know When to Fold 'em

Grade Level: Middle/Upper

MATERIALS: Decks of playing cards, graph paper (1/4", 1 cm.), scissors

ORGANIZATION: Individually

PROCEDURE: One of the more difficult skills for students is reducing fractions. This activity allows students to hone their ability to visualize how fractions reduce.

Begin by explaining that when a fraction is reduced both the number on top (numerator) and the one on the bottom (denominator) must both be divided evenly by the same number. For example, 2/4 can be reduced evenly since both the 2 and 4 can be divided evenly by 2. The same is not true for 3/4, the three is only divisible by three while the four can only be divided by itself and two.

Randomly pass out a handful of cards to each student. For the sake of this activity, give the face-cards a value that will be easy to divide (jack = 12, queen = 15, king = 18). Each student should have graph paper and access to scissors. Tell your class that they are to draw two cards from their mini-deck, using the smaller number for the numerator and the larger for the denominator. They should then cut out two sets of squares from the graph paper, one to represent the numerator and one for the denominator. For example, if 3/9 were the fraction, they would cut out two separate shapes, one of nine squares and one of three.

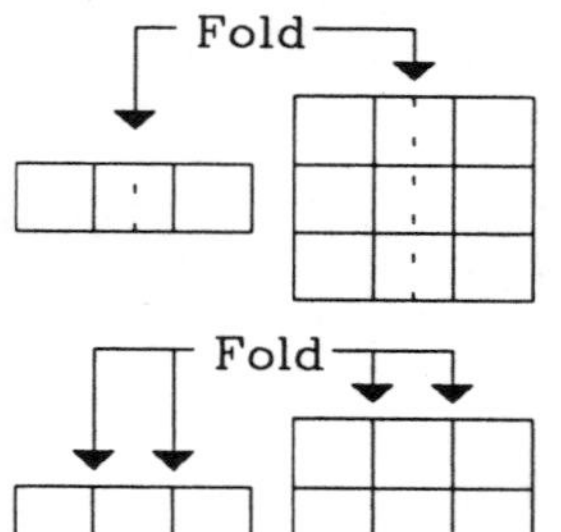

Next, have them see if the can evenly fold each of the two shapes. Have them first try to fold them in half. Obviously, there is a fold in the middle of a graphing square so this will not work. Have them try making two equal folds (into three sections). Students will find that both slips of paper can be folded into thirds (the folds align with the graph lines). This shows that 3/9 can be divided by three. The result being 1/3.

Have students continue with this experiment by drawing more fractions and cutting out and folding squares.

Hands On, Inc
2121 Rebild Drive
Solvang, CA 93463

18	Adds and Subtracts Fractions With Like and Unlike Denominators

Make a Buck

Grade Level: Middle/Upper

MATERIALS: Sets of cards with fractional parts of a dollar (Appendix G), and play money dollar bills (Appendix H)

ORGANIZATION: Groups of four

PROCEDURE: Begin this lesson by distributing the cards provided in appendix G to each group of students.

Explain that this activity will require the students to convert money amounts to fractional equivalents. Display one of the cards for the class and ask what type of fraction is represented by various coins or combinations of coins (i.e., 1/20 = nickel, 1/10 = dime, 4/10 = 4 dimes, etc.).

Give each group a supply of "dollar bills" (Appendix H) and have them place the bills in a stack in the center of the table. One student should deal five fraction cards to each student and place the remaining cards in a second pile next to the dollar pile. In turn, each student may ask any other student for a specific car. For example, "Do you have a nickel card?" or "Do you have a thirty cent card?" If that student has one, he or she gives it to the requesting player. If not, a second or third player may be asked. If none of the other players has the requested amount, a card can be drawn from the pile in the center. The next student then takes a turn.

Each time a player has a set of cards which total $1.00, the cards should be "traded in" (placed in the center pile) for a dollar bill. The fraction card pile is then reshuffled and the player may select five more cards on his next turn.

The first child with three dollar bills is the winner.

18	Adds and Subtracts Fractions With Like and Unlike Denominators

Being Shelfish
Grade Level: Upper

MATERIALS: Half-inch (1 cm) graph paper

ORGANIZATION: Individually or in groups of two

PROCEDURE: Explain to the students that they are to design a storage unit (shelving) that they would like in their bedroom at home. They must, however, follow a set of specifications for the shelves.

1. The unit is 8 feet long and 8 feet high (2 meters by 2 meters).
2. 1/2 of the space must contain rectangular spaces having a greater width than height
3. 1/4 of the space must be square shelf areas.
4. 1/4 of the space must be rectangular shelves that are greater in height than width.

Give students time to plan their shelves and remind them that there are numerous solutions. Upon completion, each student or team should share their design with the class and be able to identify 2/3, 1/4, and 1/12 of the space.

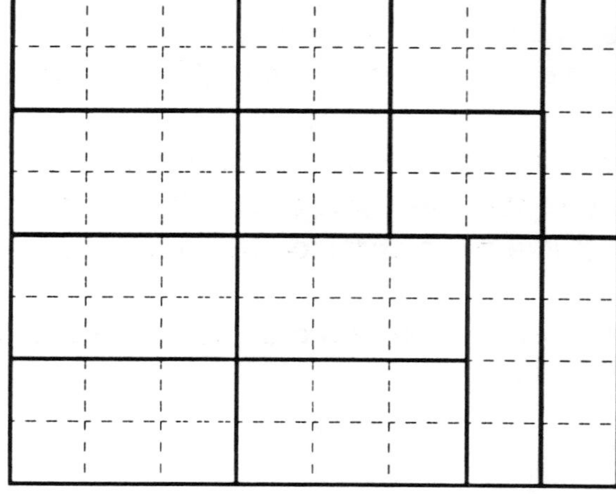

As an extension, have students make up their own guidelines for shelf configuration and have classmates draw shelves to meet these criteria. Students might even make a three dimensional model with shoeboxes or cardboard.

Hands On, Inc
2121 Rebild Drive
Solvang, CA 93463

18	Adds and Subtracts Fractions with Like and Unlike Denominators

The Magic of 60

Grade Level: Middle/Upper

MATERIALS: One-half inch graph paper (1 cm), scissors, markers

ORGANIZATION: Groups of four

PROCEDURE: This activity uses a "back door" approach to show students why they must have a common denominator to add and subtract fractions.

Distribute graph paper to each group and have them cut out 20 grids which are 10 by 6 (sixty squares) in size. We are using 60 because it has common denominators of 2, 3, 4, and 5 (among others). Do not tell this to the students.

Give them time to work together to cut from the stack of grids a representation of the following fractions: 1/2, 1/3, 2/3, 1/4, 2/4, 3/4, 1/5, 2/5, 3/5, and 4/5. You may wish to have the students label each section and color code it. You may find that this portion of the lesson will take a full class period while students figure out how to divide the sixty square grids.

Write the problem 1/2 + 1/3 on the chalkboard and have students place the 1/2 and 1/3 grids on one of the 60 grids (see below). Have them count the number of squares which are covered by the 1/2 and 1/3 representations (50 squares). Tell them that 1/2 + 1/3 = 50/60 (equalling 5/6 after being reduced). Do two or three more problems like this.

$\frac{1}{2} + \frac{1}{3}$

$\frac{50}{60} = \frac{5}{6}$

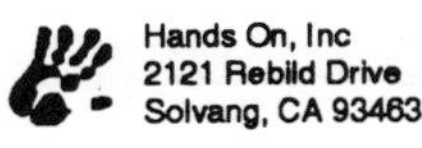

Now have them cut out several squares which are 5 by 10 (50 squares) in size. They must once again work together to divide these sections into the fractional parts listed above. They will soon find that 50 can be divided in half and into fifths, but not into thirds or fourths. Ask if they can think of any number, smaller than 60, that can be divided by 2, 3, 4, and 5. Ask if they can think of any number less than 60 that can be divided by 2, 3, and 4 (12, 24,36,48) or by 2, 4, and 5 (20 and 40).

Explain that these numbers (12, 24, 48, etc.) are common denominators and in order to add or subtract fractions the student must find a number which can be divided evenly by the denominators of both fractions.

Using graph paper squares, have students make representations of the following set of addition and subtraction equations:

$\frac{1}{3} + \frac{1}{5} = \frac{8}{15}$

$\frac{3}{4} - \frac{1}{3} = \frac{5}{12}$ 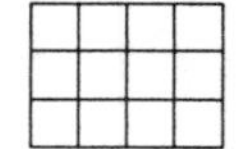

$\frac{2}{4} + \frac{1}{5} = \frac{12}{20} = \frac{3}{5}$

$\frac{1}{2} + \frac{1}{4} + \frac{1}{5} = \frac{19}{20}$ 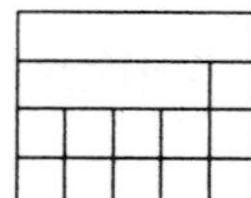

At this point, have students experiment with denominators which are greater than 5, as well as using fractions which involve mixed numbers. The process is the same and students will understand why they must find common denominators and, more specifically, what common denominators are.

Hands On, Inc
2121 Rebild Drive
Solvang, CA 93463

19	Multiplies and Divides Fractions, Mixed Numbers, and Whole Numbers

Sit Still and Learn Your Fractions

Grade Level: Middle/Upper

MATERIALS: No special materials

ORGANIZATION: Groups of ten, twelve, or eighteen

PROCEDURE: This activity utilizes teams of students to practice multiplying fractions.

Divide the class into two teams of ten, twelve or eighteen students. Teams should not have the same numbers. (These numbers are used because they have numerous factors). Team members should stand in a line facing the other team. Each student should have a classroom chair.

The activity begins with the teacher or a selected student calling out a fraction, such as, "3/4." Both teams must then spring into action to decide the number of students in the group that represents 3/4. Once this is determined, this portion of each team should sit (the remaining quarter stays standing).

The first team with the correct number of students sitting wins a point. You may wish to enforce the rule that once a student sits, he or she may not stand up again for that fraction.

Teams can use a variety of strategies to represent each fraction. After three or four practice rounds, give students three minutes to meet as a group to plan their technique.

Should a fraction be called which is not a factor of team members (3/7 for example), the team which has ALL members first sit down wins the point. If there are different numbers of students on each team, the point might be awarded to the team which can represent the fraction or you might choose to create your own set of scoring rules.

Hands On, Inc
2121 Rebild Drive
Solvang, CA 93463

19	Multiplies and Divides Fractions, Mixed Numbers, and Whole Numbers

Getting to Even
Grade Level: Upper

MATERIALS: A variety of different items for students to measure and line up end to end, rulers or centimeter sticks

ORGANIZATION: Individually or in teams of two

PROCEDURE: This lesson uses repeated addition of measurements to teach multiplication of whole numbers and fractions.

Each student should bring a set of items from home or the teacher can provide several sets of items. Some sample items would be paper clips, staples, unsharpened pencils, pieces of new chalk, etc.

Students are probably familiar with the concept of multiplication being "repeated addition." For example 3 x 4 is 4 + 4 + 4. Tell them that multiplication of whole numbers and fractions is also repeated addition. Have each student select an item to measure, they should write the actual measurement on a sheet of paper. As an example, we'll use a paper clip measuring 1 5/16." Students should first estimate how many paperclips should have to be lined up end to end in a row in order for their entire length to end exactly on an inch mark. Next, they should begin adding 1 5/16" segments while lining up paper clips. Eventually they will find that it will take 48 paper clips lined up to equal 63 inches.

From this information, have them write a multiplication equation (1 5/16" x 48 = 63") and a division equation (63" ÷ 1 5/16" = 48 paper clips). Have students write their information on a folded sheet of paper and display their equations and lines of objects for classmates to see.

Hands On, Inc
2121 Rebild Drive
Solvang, CA 93463

19	Multiplies and Divides Fractions, Mixed Numbers, and Whole Numbers

Divide and Conquer

Grade Level: Upper

MATERIALS: Pictures of money (Appendix H) of play money

ORGANIZATION: Group size to be determined by the denominator of fraction to be multiplied

PROCEDURE: In this activity, students will be dividing a given amount of money into fractional parts, dependent upon the size of the student's group.

The teacher should begin by explaining that they are going to find fractional parts of money amounts. Use the example of finding 1/8 of $10.50. Divide the class into groups of eight (the denominator) and give each group 10 one dollar bills and a fifty cent piece.

Have each group work together (without paper and pencil) to divide the money equally among all group members. To accomplish the task, the students will have the opportunity to go to the "bank" (the teacher or a designated student) to exchange dollars for coins or coins for different denominations of coins. The catch is that each group may only go to the bank twice. If, after two exchanges, they do not have the right denominations, they are eliminated from that round of play.

A second rule is that the students must distribute the money in as few coins as possible.

Once students have completed the activity select one person from each group to explain the process they used. Continue the activity with other denominations and different group sizes.

19	Multiplies and Divides Fractions, Mixed Numbers, and Whole Numbers

Times for Fractions

Grade Level: Middle/Upper

MATERIALS: Quarter inch (1/2 centimeter) graph paper, crayons

ORGANIZATION: Individually

PROCEDURE: This activity provides a visual representation of how fractions are multiplied.

Write the fraction 1/2 x 1/2 on the chalkboard and pose the question, "If you multiply two numbers, would you expect to get an answer that was larger or smaller than the factors which are multiplied?" Usually in multiplication, the product is greater; however, when multiplying fractions, the product is smaller. Tell students that today's activity will explain why this is so.

Distribute graph paper, scissors and crayons to each student and have them cut out a ten by ten grid. With one color crayon, have them color in 1/2 of the grid.

Explain that when they multiply 3 x 3, they are actually saying 3 groups OF 3 (the x symbol might be replaced by the word OF). When multiplying fractions, the same saying occurs: 1/2 groups OF 1/2. Since 1/2 is not really a group but is a part of a group, it's easier to leave this off and say, 1/2 OF 1/2.

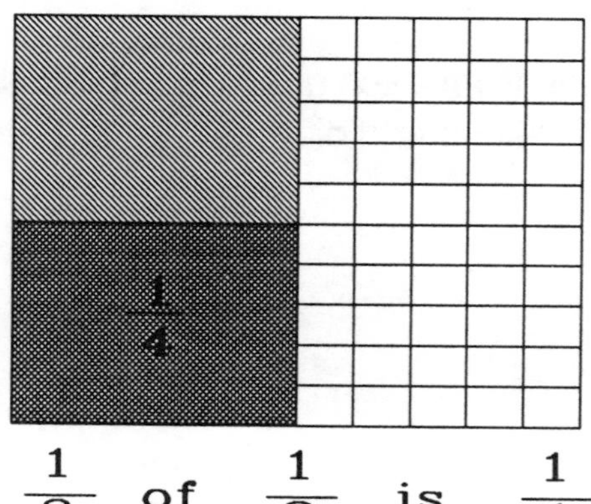

Have students use a second color crayon and color over 1/2 of the colored grid area (shown below). Ask the students what portion of the the grid is 1/2 OF 1/2 (answer: 1/4). Explain that they have just multiplied fractions.

Write some other simple examples on the board, such as 1/4 x 1/2 or 1/2 x 3/4. Have students color in the squares and explain why the answer makes sense.

19	**Multiplies and Divides Fractions, Mixed Numbers, and Whole Numbers**

Teariffic Fractions

Grade Level: Upper

MATERIALS: 12 inch (24 cm.) strips of paper, rulers, and scissors

ORGANIZATION: Individually

PROCEDURE: In this activity, students will use a trial and error approach of tearing paper to divide whole numbers into equal fractional groups.

Begin by dividing the class into groups of four and give each group fifteen to twenty strips of paper. Students should do this lesson by folding and tearing, so avoid allowing them to use rulers or scissors.

Have students begin by asking the groups to take three strips of paper and divide the strips into four equal groups. Tell them that they should accomplish this by folding and tearing the strips. Let them experiment with different strategies; however, also remind them that once they have solved this problem, they must write the fractional amount received by each student and must be prepared to explain how they got their answer.

Once students have accomplished the first part, have them do other tasks such as dividing two strips into three or five equal groups, four strips into five or six equal groups, or five strips into six or eight different groups. Each time, have students explain the fractional parts they discovered.

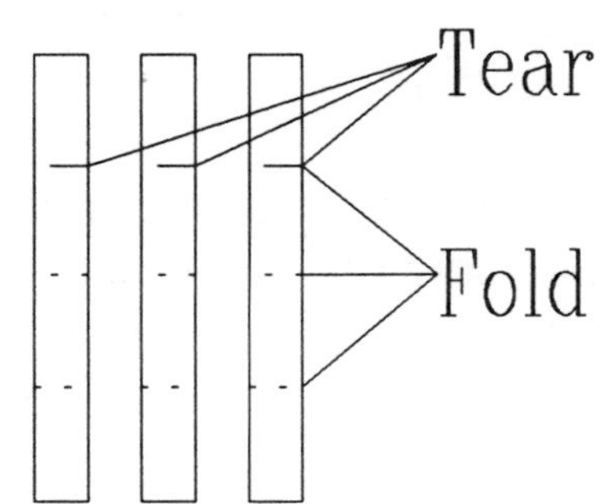

Each person receives $\frac{3}{4}$

Hands On, Inc
2121 Rebild Drive
Solvang, CA 93463

19	**Multiplies and Divides Fractions, Mixed Numbers, and Whole Numbers**

String Thing

Grade Level: Upper

MATERIALS: 4 yard strips of string (four meters), rulers, scissors

ORGANIZATION: Groups of two or three

PROCEDURE: This activity will help students understand the process of dividing fractions.

Distribute a four yard length of string to each group and ask them how many pieces they might cut which are 2/3 of a yard long. Let groups try different approaches to solving this problem.

When they arrive at an answer (6 lengths) ask each group to describe the process they used: addition, subtraction, multiplication, or division. Help them as they verbalize their responses.

Show them that they were actually dividing fractions (4 feet divided by 2/3). Write the division algorithm on the board and ask a student to come forward and explain how the process of inverting the fraction is necessary. Remind the students that multiplication and division are inverse operations and that by reversing the fraction and then multiplying, they are essentially doing the inverse of division.

Pose other situations with the string length; for example, how many pieces of string of 1/4, 1/3, or 1 1/3 of a yard could be cut form the 4 yard length. Each time relate the process to the division algorithm.

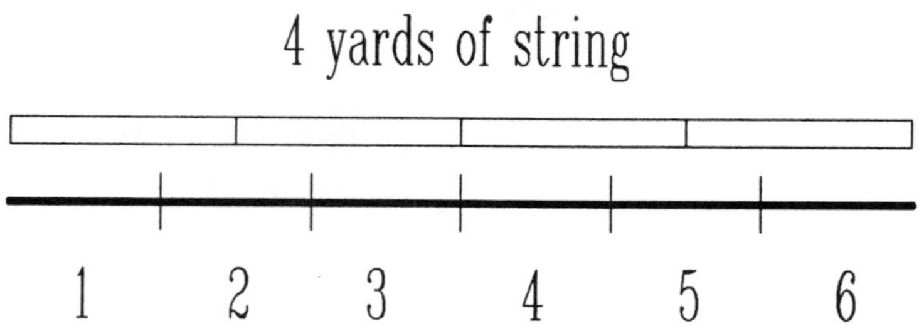

20	Multiplies Decimals

Mental Multiplication

Grade Level: Upper

MATERIALS: Four dice for each team, calculators, scratch paper

ORGANIZATION: Teams of two students

PROCEDURE: This activity provides practice in various aspects of doing mental multiplication of decimals. Some time should be spent prior to this activity discussing various approaches to mental multiplication (i.e. associative, distributive, commutative properties, and rounding numbers).

Divide the class into teams of two and give each team four number cubes. After the cubes are rolled by player 1, player 2 quickly arranges the cubes into two sets. The students should then imagine that there is a decimal point between each set of cubes and mentally multiply the numbers. They should write their answers on scratch paper.

Points are scored as follows: The student who finishes writing the answer first gets one point. Either student who gives the correct answer gets three points. If both answers are wrong, the student who is closest to the actual product gets two points. Answers should be checked on the calculator.

Students should alternate turns throwing and arranging the dice.

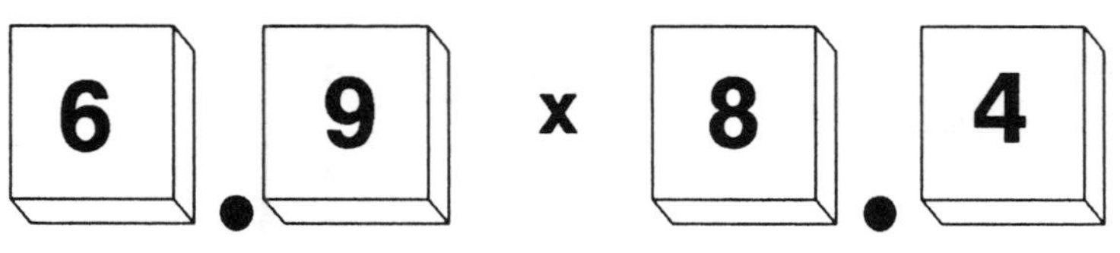

Example: Player 1 writes 56 first.
Player 2 writes 48.36.
Actual answer 57.96

Player 1 scores 1 point for being first and two points for being closest.
Player 2 scores nothing.

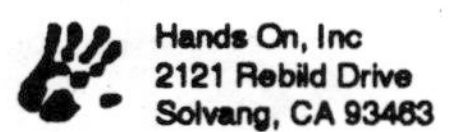

Hands On, Inc
2121 Rebild Drive
Solvang, CA 93463

20	Multiplies Decimals

Classifying the Classified
Grade Level: Upper

MATERIALS: The employment opportunities section of classified advertisements, scissors, construction paper, paste, calculators

ORGANIZATION: Cooperative groups of three or four students

PROCEDURE: The classified section of a newspaper provides numerous opportunities for students to practice practical application of the number operations. In this activity, students will search the classifieds for hourly pay rates and will then create a chart which compares several types of jobs.

Divide the class into cooperative groups and give each group the materials described above. Tell the class that they are going to make a table which compares ten different classified job offerings. They can choose to look for hourly, weekly, monthly, or yearly pay, but they must calculate the wages for each time frame for each ad.

Ads found by each group should be cut out and glued to a sheet of construction paper. They should use paper and pencil or calculators to compute the information shown in the graph below.

JOB TYPE	HOURLY PAY	WEEKLY PAY	MONTHLY PAY	YEARLY PAY

Hands On, Inc
2121 Rebild Drive
Solvang, CA 93463

20	Multiplies Decimals

Line Multiplying

Grade Level: Upper

MATERIALS: Graph paper and scissors

ORGANIZATION: Groups of two

PROCEDURE: The algorithm for multiplying decimals always emphasizes that the student must count the numbers to the right of the decimal point on the multiplicand and multiplier and then count this many numbers from the right of the product for the proper placement of the decimal point in the answer. While students do this, they seldom understand why. This lesson explains the process.

Give each team of students a sheet of graph paper and have them cut several strips of graph squares which are ten squares long. Each of these strips will represent 1 and each square will represent .1 (one tenth). Have them display a representation of the number 4.6 (as shown below). Tell them that they will now multiply 4.6 by .5 by shading squares. Using one strip at a time, have them decide, "What is .5 times (you might substitute the word OF) 1?" and "How many squares would you shade to represent this? (5). They should shade the first four strips as shown

Next, ask, what is .5 times (of) .6?" They should shade 3 of the six squares (as shown). Ask students to count the numbers of squares that are shaded (23) and ask how much each square represents (.1 or 1/10). Elicit from the class that 23 tenths should be written as 2.3.

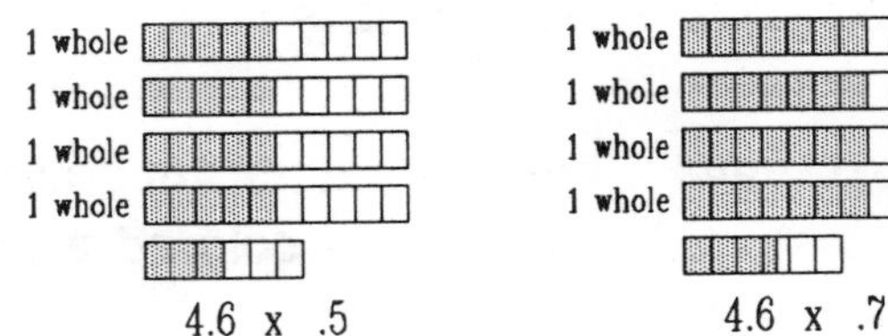

The teacher will probably have to explain this process two or three times before the "AH HA" response, but be patient. Once students understand the concept of multiplying by .5, have them do the same problem multiplying by .7. All is the same except that students will have to estimate .6 x .7. Eventually, point out that 6 x 7 is 42 so 4.2 squares would have to be colored. Give students several examples to work with. In time, they will see why the decimal point cannot simply be brought straight down as it is in addition and subtraction of decimals.

Hands On, Inc
2121 Rebild Drive
Solvang, CA 93463

20	Multiplies Decimals

Small Times for Decimals
Grade Level: Upper

MATERIALS: Quarter inch graph paper, scissors, and crayons

ORGANIZATION: Whole class activity

PROCEDURE: This activity gives students a visual representation of how decimals are multiplied.

Begin the activity by writing .3, .27, and .9 on the chalkboard and select a student to come forward to rewrite the three numbers in order from least to greatest (.27, .3, .9).

Next, ask students, "if two of these numbers are factors and one is a product, how would you write this sentence?" (.3 x .9 = .27). Now ask how it is possible to multiply two numbers together and get an answer which is SMALLER than the factors which were multiplied. Tell students that in today's lesson, they will see how this can happen.

Distribute graph paper, scissors, and crayons and have students cut out a 10 by 10 grid. Have them use a crayon to color .3 of the grid (30 squares). Remind students that when they multiply 4 x 3, they are actually making 4 groups of 3. When they multiply .3 x .9, they are actually finding 3 tenths of a group of 9 tenths. The OF can (and should) be substituted for the "x" symbol.

Next, have them use a different crayon to color .9 (9 tenths) of the previously colored area (27 squares). Show them that .3 x .9 = .27.

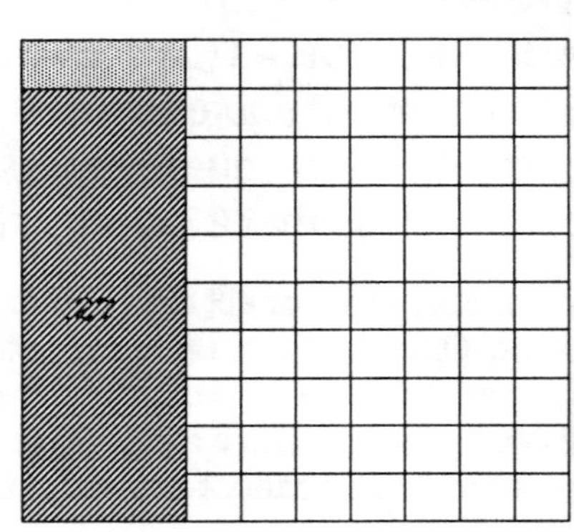

Give students other simple decimals to multiply and select students to come forward and explain why products in multiplication of decimals are smaller than the factors.

Hands On, Inc
2121 Rebild Drive
Solvang, CA 93463

21 Divides Decimals and Computes and Rounds Remainders

My Calculator Remains

Grade Level: Upper

MATERIALS: Calculators for each student

ORGANIZATION: Individually

PROCEDURE: Students are usually taught to write remainders when doing the division algorithm. Calculators, on the other hand, present "remainders" as decimal equivalents. This activity helps students to understand the difference between whole number remainders and decimal equivalents.

Write the division problem $3\overline{)13}$ on the chalkboard and ask students for the quotient. Most students will be able to respond "4 remainder 1." Discuss the meaning of the word remainder and be certain the students understand that since 3 will fit evenly into 12 four times, there would be 1 left over or remaining if thirteen items were divided into 3 equal groups.

Have them do this same problem on their calculators. The answer they will get is 4.3333333... Ask the students to describe what this means. Eventually, elicit the response that instead of placing 1 "extra" to the side, as in remainder 1, the calculator also divides the remainder 1, evenly, into 3 equal parts. Ask students how they could round .3333333... to the nearest 100th and how could they write it as a fraction (.33 and 1/3).

Give students some time to think about thow they might be able to convert this decimal into a whole number remainder as they did with their paper and pencil approach. Give them the hint that .33 or 1/3 means that they have DIVIDED remainder 1 into 3 (divisor) equal parts.

The solution is that students should multiply .33 x 3, equaling .99, and would then round off to 1. Explain that the reason this does not fall evenly on 1 is that they rounded the .33. Present other division problems with different divisors. Students will learn that by multiplying the decimal equivalent by the divisor, they will find the whole number remainder.

Hands On, Inc
2121 Rebild Drive
Solvang, CA 93463

21	Divides Decimals and Computes and Rounds Remainders

Performing Surgery on Decimals

Grade Level: Upper

MATERIALS: Ten inch or centimeter strips (as shown below) scissors

ORGANIZATION: Individually or in teams of two

PROCEDURE: This is a beginning lesson in demonstrating to students how decimals are divided.

Begin by writing the algorithm 3 $\overline{)2.1}$ on the chalkboard. Ask for a volunteer to estimate the answer. Most students will respond "7," which is the correct digit but ask if it makes sense that 2 and 1 tenth, when divided into 3 equal groups can have seven 7 "wholes" in each group. The students will see that the answer must be .7. Then pose the question, "What is 7 tenths of a group?" This lesson will help students visualize this.

Distribute strips of paper as described. Each strip should be divided into ten equal sections. Write 3$\overline{)1.7}$ on the chalkbaord and have students figure out how to divide 1 and 7 tenths strips into three equal groups. You may wish to have them work in teams or individually.

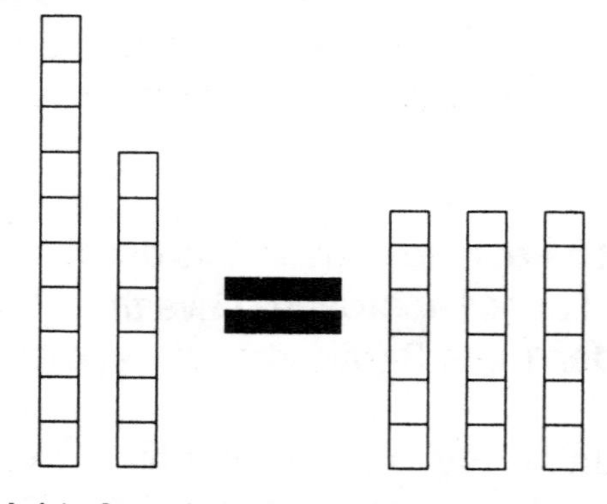

1.7 divided in 3 equal groups .567 per group

Students will take some time to do their cutting and figuring, but generally this will not be too difficult for students to master. However, as students visualize the process of dividing a decimal number into groups, following the division algorithim "rule" of "bring the decimal point straight up when dividing" will make sense to them.

Repeat the process with other examples

Hands On, Inc
2121 Rebild Drive
Solvang, CA 93463

22 | Identifies Equivalent Fractions, Decimals, and Percents

Fractions on My Hands

Grade Level: Middle/Upper

MATERIALS: String and straws

ORGANIZATION: Individually or teams of two students

PROCEDURE: One way to teach students the concepts of percentage, decimals and fractions is by relating them to money. In this activity fractional parts are related to a different fundamental understanding, the clock face.

Give each student a length of string, long enough to tie end to end and make a circle on the desktop. This will be the clock face. Give the students two straws and have them cut a portion off of one of them to make an hour hand and minute hand. Using these hands, students will mark off a section of the "whole" clock face thus creating a fractional relationship.

Begin by having the students arrange the hands of the clock to demonstrate the fraction 1/2. Ask what time is represented by this (there are numerous answers, including 12:30, 6:00, 7:05, 1:35, etc.). Now ask students to arrange the hands to show 50% and .5 of an hour (the same configurations apply).

Once students understand that the hands can represent fractions, decimals, and percentages, call on students to suggest numbers for classmates to demonstrate. In each class, call on other students to give the fractional, decimal, or percentage equivalent of the number called out by the student. The examples below show several possible combinations.

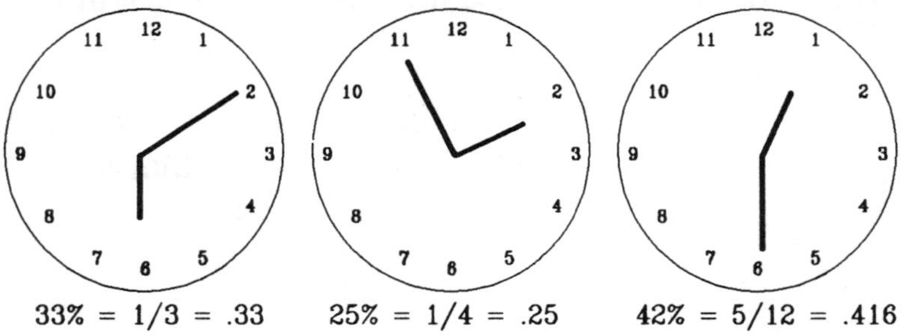

Hands On, Inc
2121 Rebild Drive
Solvang, CA 93463

22	Identifies Equivalent Fractions, Decimals, and Percents

Seeing is Receiving

Grade Level: Middle/Upper

MATERIALS: Different colors of construction paper, rulers, scissors, and markers

ORGANIZATION: Individually as a whole class or in groups of two

PROCEDURE: This activity gives students practice in finding equivalent fractions.

Begin by cutting several long strips of paper approximately 1/2 inch (2 centimeters) wide. These strips will be used to show equivalent fractions, so their overall length will be determined by the fractions chosen for a particular lesson. We'll use the example of finding equivalent fractions for 12.

Have students cut six 12 inch strips. Have them measure, cut and label the following pieces:

- One strip cut into 2 pieces, each 6 inches long and labeled 1/2.
- One strip cut into 3 pieces, each 4 inches long and labeled 1/3.
- One strip cut into 4 pieces, each 3 inches long and labeled 1/4.
- One strip cut into 6 pieces, each 2 inches long and labeled 1/6.
- One strip cut into 12 pieces, each 1 inch long and labeled 1/12.
- One strip should remain uncut.

Have students place the strips as shown. Then have them record at least five equivalent fractions which they see, for example, 1/6 = 2/12, 1/4 = 3/12, etc. Discuss pieces which have no equivalents (1/4 cannot be expressed as 6ths or 3rds). Repeat the process with several different length strips.

1/2 | 1/2
1/3 | 1/3 | 1/3
1/4 | 1/4 | 1/4 | 1/4
1/6 | 1/6 | 1/6 | 1/6 | 1/6 | 1/6
1/12 | 1/12 | 1/12 | 1/12 | 1/12 | 1/12 | 1/12 | 1/12 | 1/12 | 1/12 | 1/12 | 1/12

If students have difficulty, have them place strips on top of one another (i.e., 2/4 on top of 1/2) to see equivalencies.

Hands On, Inc
2121 Rebild Drive
Solvang, CA 93463

22	Identifies Equivalent Fractions, Decimals, and Percentage

Fractional Cents

Grade Level: Upper

MATERIALS: Velcro strip or two sided tape, large copies of pennies, nickels, dimes, quarters, half-dollars, and dollar bills.

ORGANIZATION: Whole class activity

PROCEDURE: Fractions, decimals, percents, and money amounts share a trait of being a "part" of a larger unit. For example, 1/2, .5, 50%, and $.50 all represent one-half of a whole unit. This activity uses dollar amounts as a means of showing students these relationships.

In preparation attach a long strip of Velcro across the top of the chalkboard. You will find that this Velcro strip is useful for display in many classroom activities and can be purchased at hardware stores or lumberyards. Double sided tape will also work; however, it is not nearly as permanent. Duplicate and cut out several large copies of coins and dollar bills. Place Velcro tabs on the backs of these cut-outs.

Begin the lesson by posting several coins on the Velcro strip. Ask students to record the money amount, the decimal equivalent, the percent, and the fractional amount of $1.00 represented by the coins. Continue this process with many different sets of coins, eventually using dollar bills (mixed numbers and percentages greater than 100%).

As an extension, you can change the designation for the "100% amount." For example, you might tell students that, "$2.00 is the amount which is 100% of the total. What part of $2.00 is $.50?" (1/4, .25, or 25%).

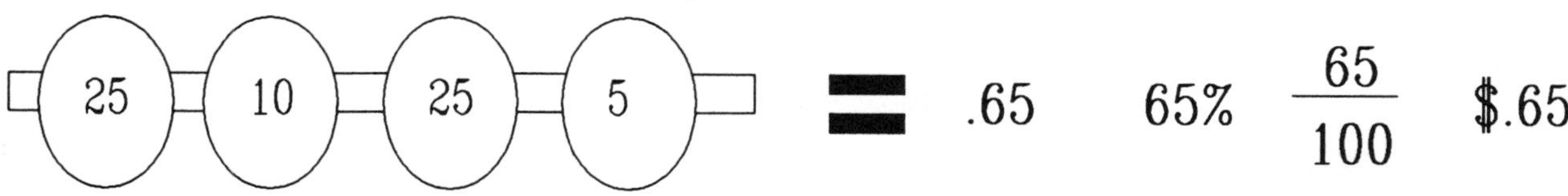

Hands On, Inc
2121 Rebild Drive
Solvang, CA 93463

22	**Identifies Equivalent Fractions, Decimals, and Percents**

I've Got Your Number

Grade Level: Upper

MATERIALS: One 3" x 5" card for each student

ORGANIZATION: Whole class activity

PROCEDURE: In this activity, students will use their telephone numbers to generate the numerators and denominators of fractions.

Each student should write their phone number (or a made up phone number) on a 3" x 5" card. The cards are passed around the room for each student to make a list of class phone numbers. Individual students should then make fractions from the phone numbers, adding the first three numbers for the denominator and adding the last four digits for the numerator.

An example would be 344-8332. The fraction formed by this number would be 16/11 (8 + 3 + 3 + 2 = 16 and 3 + 4 + 4 = 11). The student then tries to find another phone number fraction that when added to or subtracted from the first fraction makes a whole number.

For example, the number 344-2112 creates the fraction 6/11. 6/11 plus 16/11 equals 22/11 which can be renamed as 2. These two phone numbers would be considered a "match" by the students.

Have students find as many matches in their lists as possible.

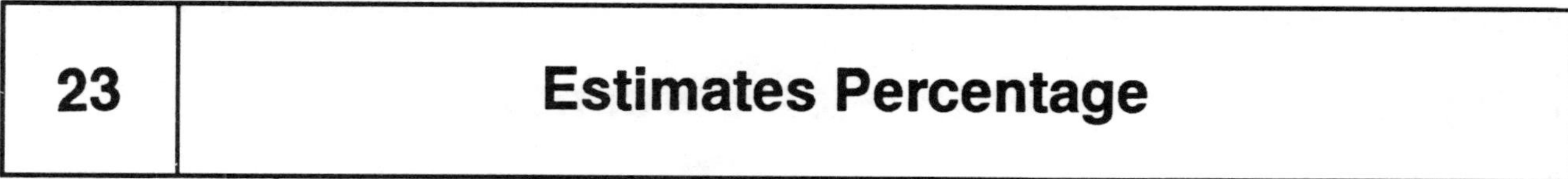

23 Estimates Percentage

A Hill o' Beans

Grade Level: Upper

MATERIALS: Large beans painted on one side and natural on the other

ORGANIZATION: Teams of two

PROCEDURE: This lesson uses ratios of one color bean to another to allow students to estimate percentages of a given number. It also provides practice in letting students recognize that percentages have a constant fractional equivalent.

Divide the class into teams of two students and give each team fifty or so beans as described above. You will need to do some modeling of the activity for the students before they are able to work on their own.

As a class, have students make an array to represent the number 24 (see below). Any array will work. Tell students that you want them to turn over 100% of the beans in their arrays so that only one color shows. Next, ask students to turn over 50% of the beans and then ask them to tell the ratio of beans that were turned (1 out of every 2 beans). Continue with this process, turning 25% (1 out of 4 or 1/4), and then 75% (3 out of 4 or 3/4). Over a period of a few days, include all basic fractions with denominators of 2, 3, 4, 5, 8, 10, and 12.

Once they are familiar with thei concept have one member of each team take a pile of beans and "roll" them. The other student should first estimate the percent of beans, color-side up, and then sort to check this estimation.

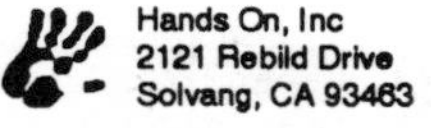

Hands On, Inc
2121 Rebild Drive
Solvang, CA 93463

23	Estimates Percentage

Parts is Parts

Grade Level: Upper

MATERIALS: Graph paper and pencils, magazines with animal pictures

ORGANIZATION: Individually or in groups of two

PROCEDURE: This is an unusual lesson in which students will estimate the percentage of an animal's head, neck, body, and legs. Students will use graph paper to do their estimating.

Have students find side-view pictures of two or three animals. Pictures must be small enough (and large enough) to fill at least half a sheet of graph paper. Have them cut out a silhouette of an animal and trace it onto the graph paper.

Have students estimate, and then count, the total number of squares covered by the animal. They should then estimate/count the number of squares covered by the head, the neck, the body, and the legs separately. Students should then compute the percentage of each section and make a poster as shown below.

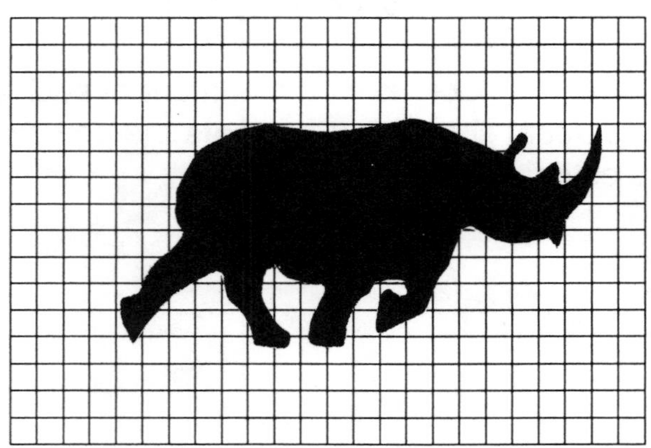

Hands On, Inc
2121 Rebild Drive
Solvang, CA 93463

23	Estimates Percentage

Seeing Stars

Grade Level: Upper

MATERIALS: Graph paper, an American Flag, copies of the American Flag (Appendix I)

ORGANIZATION: Individually or in cooperative groups

PROCEDURE: This is an unusual approach to finding percent in which students will use various methods to determine what percentage of an American Flag is red, white or blue, and what percentage is made of stars.

Display a flag for students and have them write down their estimates as to what percentage of the surface area of a flag is red, white, or blue and what percentage is made up of stars. Next, distribute graph paper and copies of Appendix A to the students.

Have students trace the flag onto the graph paper. Tell students they should use the squares of the graph paper to help them correct the percentages they estimated. There are several approaches that they might use and remind them, that upon completion, one group member will need to share the process used by his or her group. In any case, each group must write down the percentages for each category and share their data with the class.

You may wish to add an unusual "twist" to the exercise by reducing or enlarging Appendix A when it is copied, or by distributing different size graph paper to each group. As an extension, you can have students select a flag from another country and compute its distribution of colors.

23	Estimates Percentage

A-maze-ing Per-Sense

Grade Level: Upper

MATERIALS: Graph paper and markers

ORGANIZATION: Teams of two

PROCEDURE: This activity is aimed at helping students develop skill in estimating percentages.

Begin by explaining that the term "per cent" is Latin meaning "for each one hundred;" when we say 50% we mean 50 for each 100 — or one half. Next, have students outline a ten by ten section of graph paper and, using a heavy marker, have them draw a simple maze (a sample is shown below).

Explain that they are going to play a game where they will have to steer a special spacecraft through a maze. In order to steer through the maze, the "pilot" will have to give directions in terms of power for a move of 6 squares. Once the students understand the procedure, have them create mazes for their partners.

When the mazes are complete, the partner is to steer through the maze but may only see it twice -- once to start and once halfway through. The pilot will give directions through the maze to the partner. The directions should be given as percentages of power (each square equaling one percent) and the compass direction. The partner should draw his path on a blank sheet of graph paper. After completion, the blank should be filled in and matched with the original.

The students can race each other, but for the first few times the mazes should have some restrictions, so that they are not too treacherous (i.e., walls must be at least five squares long, no dead ends, etc.).

Hands On, Inc
2121 Rebild Drive
Solvang, CA 93463

24	Computes Percent of a Number

Percentage as a Ratio
Grade Level: Upper

MATERIALS: Graph paper and scissors

ORGANIZATION: Cooperative groups of three or four

PROCEDURE: One concept which students must understand when working with percentage is that a percent represents a fractional part or relationship. In this activity, students will use graph paper to show how ratio and percent relate.

Write "13% of 52 is ?" on the chalkboard. As with every example you give students, first ask the students to estimate a reasonable answer. Even if students know how to do the computation, explain that in this lesson they are going to solve this equation using graph paper to create a ratio.

Divide the class into groups and distribute graph paper. Have them count and cut a thin strip of graph paper 52 squares long. Number each square. Ask them to cut a second strip of plain paper which is 100% as long as the first (let them figure out that it too will be 52 squares long).

Have them fold the second strip in half and ask what percentage is represented by the fold (50%). Lay it parallel to the first strip and ask what 50% of 52 would be. Fold the second strip in half again. What percentages would be represented by the two new folds (25% and 75%)? With this knowledge, have students work together to figure out how to create a fold at 13% so they can align the two strips to find 13% of 52.

Have each group share their method with the class and then, using these two strips of paper, pose different percentage amounts of 52 for the students to find.

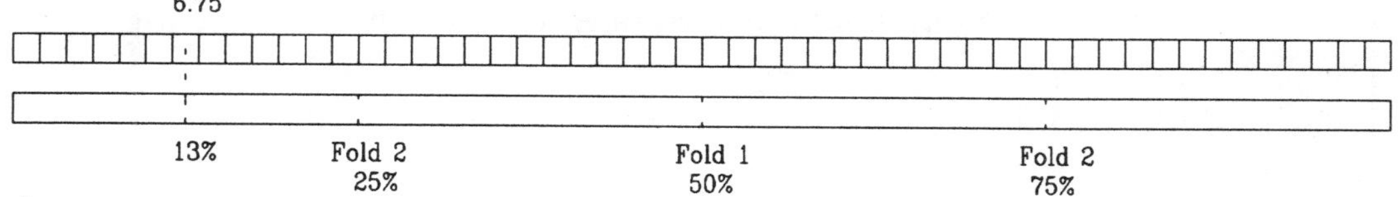

Hands On, Inc
2121 Rebild Drive
Solvang, CA 93463

24 Computes Percent of a Number

A Relative Fraction

Grade Level: Upper

MATERIALS: Graph paper and pencils

ORGANIZATION: Individually

PROCEDURE: This activity gives students a different approach to finding the percent of a given number and helps explain the concept that percentage is based upon the fractional parts.

Give students a sheet of graph paper and ask them to outline a grid which would represent the number 25 (a five by five area). Have them explain how they could figure out how many squares should be shaded to represent 20% of 25. Although some students may be able to figure this out, explain that you are going to show them a different way to approach this.

First, ask the students how many squares would have to be shaded to represent 100% of the grid (all of them). Explain that this establishes a ratio. In other words, if 100% = 25, there is a ratio of 4 to 1 (4% is represented by each square). Based on this ratio, ask how many squares would need to be shaded to represent 8% of 25 (2), 16% of 25 (4), or 60% of 25 (15). Give the students many opportunities to respond, as this is the key point of this lesson. Next, return to the original request to have students shade 20% of the grid (5 squares).

When students understand, extend the lesson to include equivalents of half or a quarter of a square. For example, 10% would be represented by shading 2 and 1/2 squares.

Have them make different sized grids to work with, but at this point limit these to numbers which divide evenly into 100.

Hands On, Inc
2121 Rebild Drive
Solvang, CA 93463

24	Computes Percent of a Number

A Tip for Practicing Percents

Grade Level: Upper

MATERIALS: No special materials are necessary

ORGANIZATION: Whole class

PROCEDURE: This is teacher led activity which can be ongoing over several days or weeks. In this activity, students will have to mentally perform percentage reductions and increases.

Explain to your class that there will numerous times in their lives when they will have to figure a percentage discount or percentage increase, but will not have the use of a calculator or paper and pencil. They will need to do the computation mentally.

Begin with a discussion of 100% increases and discount. Explain that this has the effect of "doubling" or "eliminating" an original number. For example, if the price of an item were $12.00, a 100% increase would make the cost $24.00; a 100% decrease would make the cost $0.00. Throughout the day, select students to respond to the teacher's questions such as, "The cost is $5.00, what's a 100% increase?"

On the second day introduce the concept of a 50% increase and decrease and explain that this is either half- again as much or half less. Continue quizzing the students during those in between moments in class.

The third day should bring a discussion of a 10% increase and decrease. This will certainly be more difficult and may take several days of practice. Be certain that students understand that 10% is equal to 1 out of every 10 parts of the original. For example, 10% of $12.00 would be $1.20, or $.12 for each dollar (also point out the movement of the decimal point one place to the left). The $1.20 would then be added to the $12.00 to find a 10% increase ($13.20).

Eventually, have students try more difficult percentages such as 20% or 30%. It's easiest to teach these as 10% doubled or tripled. The final step would be 15% (tips) in which they take 10%, halve this amount (5%) and add the two together.

Hands On, Inc
2121 Rebild Drive
Solvang, CA 93463

24	Computes Percent of a Number

Top Ten

Grade Level: Upper

MATERIALS: Paper, ruler, crayons, and pencils

ORGANIZATION: Groups of two

PROCEDURE: This activity is one in which students will compute and display percentages of student preferences.

The teacher should assign each team a topic of interest for students to survey. For example, favorite sports, music groups, songs, foods, etc. Each group needs to make a sign-up sheet for each choice in their survey.

Sign-up sheets should be placed around the room and each student should sign one choice from each topic. When all students have made their choices, each group should then compute the percentage of the class which selected each item. Divide the number of students into 100. Use this number and multiply by the number of votes to get a class percentage (e.g. 100 divided by 25 = 4; 4 x 6 = 24%).

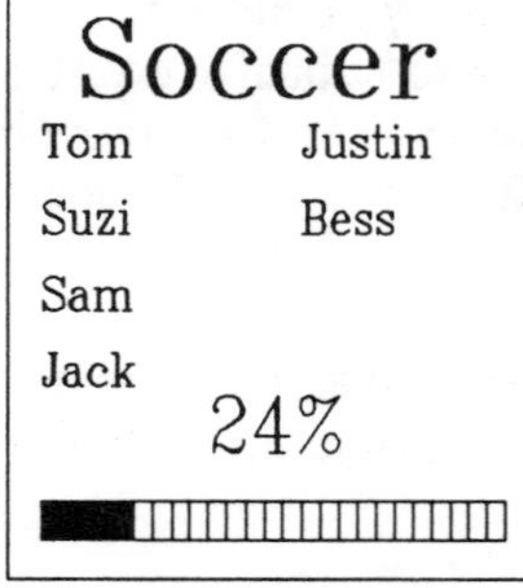

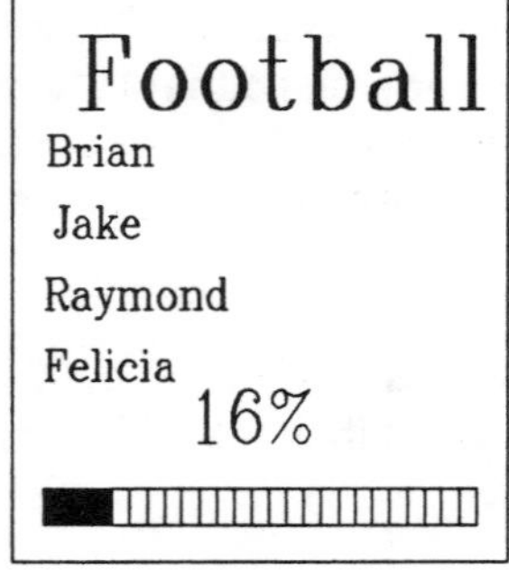

Surfing
Ruben
Travis
Billy
Annie
Joe
Laura
Scott
28%

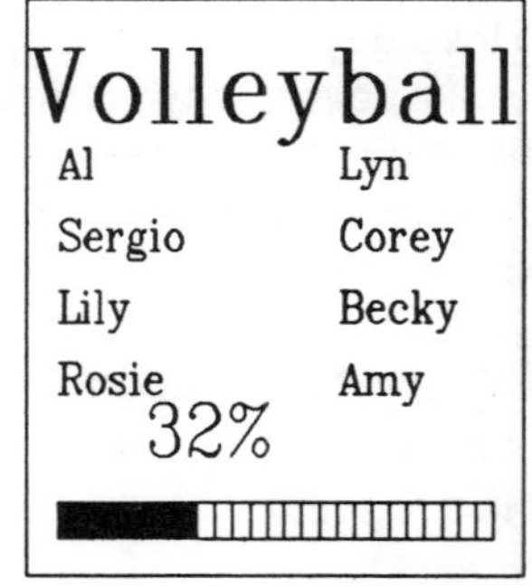

25 Computes Percentage Relationships Between Two Numbers

Picture This

Grade Level: Upper

MATERIALS: Graph paper

ORGANIZATION: Individually

PROCEDURE: This activity gives students practice in estimating percentage relationships between numbers by drawing pictures to represent the percentage.

As mentioned earlier, there are three different scenarios in which students compute percentage. This activity focuses on finding a percent relationship between two numbers, for example, 24 is what percent of 42?

Given this example, distribute graph paper and have students outline an area which would include 42 squares. Next, have them shade 24 of these squares. Using this picture, have them estimate the actual percentage required in this problem.

As you ask for student responses, guide the discussion with questions such as, "Is more or less than half of the area shaded? If more than half is shaded, would this mean more or less than 50% is shaded? What percent of your picture is shaded? What percent of your picture is unshaded? If the unshaded is less than 50%, would the shaded have to be greater than 50%? Etc."

Provide several examples for students to draw, each time stretching the conversation to include discussion of the difference between 25% and 33% or 67% and 75%. Also relate percentage to fractional amounts so students understand that percents and fractions both describe the same parts.

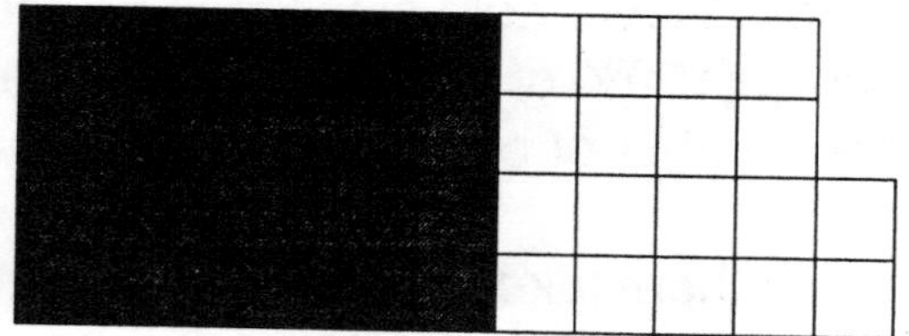

25 Computes Percentage Relationships Between Two Numbers

Bean Sense

Grade Level: Upper

MATERIALS: 100 beans for each student (or team), a paper cup

ORGANIZATION: Individually or in teams of two

PROCEDURE: A fundamental understanding which students must grasp is that 100% is equal to 1 whole (or complete set) and that any lesser percent is a fractional portion. This activity, combined with the next activity teach the concept of 100%, how you can have more than 100%, and how less than 100% is a fractional part of a whole. It may seem absurd in its simplicity but it is essential to understanding percentage.

Give each student or team a handful of beans and have them count out 100 of them. Tell students that they will begin the lesson by working with the 100 "pieces of candy." They will eventually work with different sized sets. Have them place the 100 beans (candies) in a paper cup and tell them that the cup will be called the candy dish and will be equal to 1 whole unit.

Have them remove 5 candies and ask, "What percent of the candies did you remove?" Accept answers of 5%, 5/100, 5 out of 100, etc. Ask them if they removed the whole bowl (no), and if they removed five whole candies (yes). Emphasize the terminology that the candies, even though they are complete units unto themselves are only part of the candy dish. In other words, they are a fraction or a portion of the whole. Discuss this and elicit student explanations in their own words. Have them place the five candies back in the bowl and ask what percent is now in the bowl (100% or one whole). Repeat this process several times, each time removing a different number of candies until all students understand.

Next, have them take two more beans out of the extra pile and place them in the bowl. Again quiz them about percents; What percent of candies are now in the bowl (102%)? Is 102% more than one unit? Is 102% less than 2 units? Etc. Repeat the process with adding and removing candies.

Hands On, Inc
2121 Rebild Drive
Solvang, CA 93463

25	Computes Percentage Relationships Between Two Numbers

Bean Sense, Too

Grade Level: Upper

MATERIALS: 100 beans for each student or team, paper cups

ORGANIZATION: Individually or in teams of two

PROCEDURE: This lesson is a continuation of "Bean Sense" on the previous page. In this activity, students will learn the process for computing the percentage relationship between two numbers.

Have students place 100 beans in a paper cup. This will be their "candy dish." Remind them that in the previous lesson, the candy dish is equivalent to 1 whole unit and that each individual candy represents 1% (a fractional part of the whole).

Now, have students remove 20 beans and put them aside and tell them that the candy dish "whole" (100%) is now equal to 80 beans. Have them remove 5 beans and ask, "What percent of the beans have you removed?" Undoubtedly, many students will respond 5%. Remind students that 5% of 100 is 5 candies; therefore, 5% of 80 cannot also be 5. To clarify, write

5 out of 100	5 out of 80
$\frac{5}{100}$	$\frac{5}{80}$

Ask the students the answer to 5/100 (5%) and let them discuss how this might have been computed (5 divided by 100 = .05. (= 5%). How would they compute 5/80? (5 divided by 80 = .04 = 4%).

82 out of 80
$\frac{82}{80}$

Have the students place the five candies back in the bowl and ask what percent is now represented (100%). From the extra pile, have them add 2 more candies and ask if the now have more or less than 100% (more). Have them represent this number as a fraction and once again ask how they would find the percent relationship (82 divided by 80 = 1.02 = 102%).

Pose other problems using the candy dish approach, emphasizing that the student needs to evaluate the problem as to what makes sense -- a percentage greater than 100% or less than 100% (a fractional part of the whole).

Hands On, Inc
2121 Rebild Drive
Solvang, CA 93463

25 Computes Percentage Relationships Between Two Numbers

Bean Taxes and Bean Discounts

Grade Level: Upper

MATERIALS: 100+ beans for each student or team

ORGANIZATION: Individually or in teams of two

PROCEDURE: The two previous lessons introduced the concept of finding a percentage relationship between two numbers. This lesson uses the same approach of manipulating beans to show students how to compute percentage increases (as in taxes and tips) and decreases (as in sale discounts).

Distribute beans to each student or team and have them count out 100 beans to place in a pile. This pile will represent 100% (one whole unit). From a separate pile, have the students get 6 beans. Ask them what percent of the whole unit of beans this would represent (6%). Also elicit the number of beans that would be present in the 100 pile if there had been a 6% discount in beans (94) or if there had been a 6% sales tax (106). These are simple computations when working with 100, but now have students create a pile of 75 beans which will represent 100%.

Have students remove 6% of these beans and tell how many would remain in the pile. They will find that the task becomes more difficult. Tell students that they must follow the same procedure as when they were working with 100 beans; that is, 6% of 100 = 94 beans. To do this they used a two-step process: .06 x 100 beans = 6 beans; AND 100 beans - 6 beans = 94 beans. See if any students can figure out how to do this computation as a one-step process (if 6% were removed, 94% would remain; therefore, .94 x 100 beans = 94 beans).

In the situation of 75 beans, the same logic still applies. If 6% were removed, 94% would still remain; therefore, .94 x 75 beans = 70.5 beans. Have them do the same process with an increase of 6%. If 6% were added, 100% would remain, PLUS the extra 6% would mean that 75 beans x 1.06 = 79.5 beans.

Present several similar problems for students. As they do the computation, be certain to manipulate the beans to have a concrete example of what is occurring as they do their multiplication.

26	Computes Rates, Ratios, Proportion, Scale, and Percentage

Ratios, Old Bean!

Grade Level: Upper

MATERIALS: Red beans, white beans, and various containers

ORGANIZATION: Individually

PROCEDURE: This is a beginning lesson in which students will discover the concept of proper and improper fractions functioning as a ratio. In this case, the ratio between red and white beans.

The teacher should prepare for the lesson by placing a variety of red and white beans so that each student can work independently.

The student should first look at the container and estimate the number of red and white beans. They should then make an estimate as to how many reds there are for each white, or vice versa. They should record their responses.

Students should then actually count the beans and write the ratio as both a proper and improper fraction in its lowest terms (reduced).

Next, have students write a brief paragraph in which they tell about two items in real life which would probably have the same ratio they discovered with the beans. For example, a ratio of 1/4 would be the ratio of horses' heads to their feet; a ratio of 4/1 might represent the ratio of quarters to one dollar.

Have students share their ideas with the class.

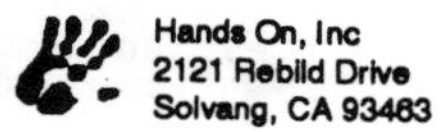

Hands On, Inc
2121 Rebild Drive
Solvang, CA 93463

26	Computes Rates, Ratios, Proportion, Scale, and Percentage

Salt Water Ratios
Grade Level: Upper

MATERIALS: Glass containers, teaspoons, measuring cups, water, salt, and masking tape

ORGANIZATION: Groups of two

PROCEDURE: This activity combines a science lesson with students determining a ratio (fractional relationship) of salt to water through the process of evaporation. It will require several days to complete.

Divide the class into pairs and give each team a water container. Using masking tape, they should write their names to label their container. Have them measure and pour 1 cup of water into the container and then measure a number of teaspoonfuls of salt into the container. They should record the number of teaspoons used and compute the ratio of salt to water (1 cup of water = 64 teaspoons). The teacher should write a list of these ratios but keep it secret from the class. Students should then stir the water to be certain that all the salt is dissolved.

Place the containers in a location where the water can evaporate naturally.

After the water has evaporated, distribute the glass jars (now containing only salt) to students. They should not have their original jars. Have them measure the salt using teaspoons and once again compute a ratio. Using the list, the teacher should call on students to see if the ratios match those on the teacher's list.

As an extension, you might have students write the ratios as percentages or decimals and have them explain how fractions, decimals, percents, and ratios are similar.

Hands On, Inc
2121 Rebild Drive
Solvang, CA 93463

26	Computes Rates, Rates, Ratios, Proportion, Scale, and Percentage

Shrink to Fit

Grade Level: Upper

MATERIALS: Scissors, 1/4"graph (1/2 cm), blank paper, straight edge, markers or crayons

ORGANIZATION: Individually

PROCEDURE: In this activity, students will practice their ability to work with ratio and percent and will see how this ability is used in a "real-life" situation.

Begin by explaining that in magazines and newpapers certain jobs require the ability to shrink or enlarge photos and advertisements to fit a given size. Tell your students that each group is going to create a folder of advertisements which will then be sized to fit a class magazine.

Have students decide what they want to advertise. Have them create their advertisement on a piece of graph paper. The ad need not fill up the whole page and should be simple.

When ads are complete, have students trade within their group. Individual students should decide the amount by which they will reduce their original (by 1/2, 1/3, or 1/4). If the reduction number was 1/3, they would divide the graph tracing into thirds. To do this they will need to make a box on a separate piece of blank paper that is one-third of the original ad (see diagram). They will then need to grid that box with the same number of lines as the framed ad. Once this is done they should retrace the ad into the smaller box as precisely as possible.

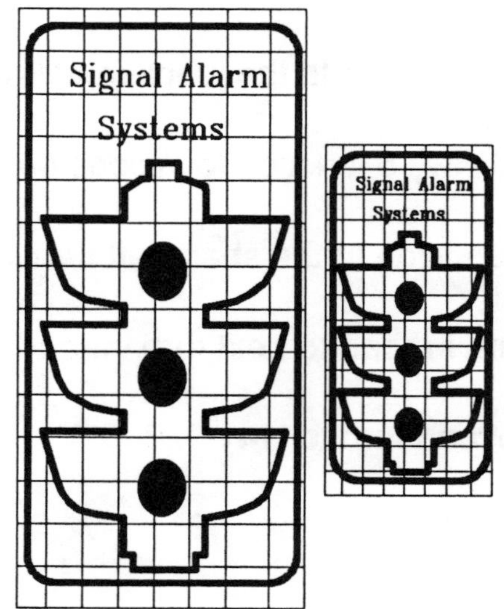

Once the ads have been re-sized they may be returned to the owner/advertiser to color in and detail as desired. These final ads can be displayed and even made into a mock publication with fake type set into the margins.

26	Computes Rates, Ratios, Proportion, Scale, and Percentage

This Map Rules
Grade Level: Upper

MATERIALS: Road maps, rulers, calculators

ORGANIZATION: Teams of two or more

PROCEDURE: This activity uses a road map scale as a means of having students identify scale and practice multiplying by fractions.

Begin by distributing road maps to each group of students. The students should study the map's mileage scale and estimate the mileage between two points. Thye should then measure with a ruler. Once measurements have been made, have students write multiplication equations for map distances.

For example, given a map with a scale of 1" = 18 miles, the distance from Green River to Price might be broken down and computed in smaller measurements such as:

Green River to the junction of Hwy 6: 5/8"	$\frac{5}{8} \times 18 = 11\frac{1}{4}$ miles
I70/Hwy 6 junction to Woodside: 1 1/8"	$1\frac{1}{8} \times 18 = 20\frac{1}{4}$ miles
Woodside to the sharp left turn in the road: 1/2"	$\frac{1}{2} \times 18 = 9$ miles
Turn in road to next turn in road: 3/16"	$\frac{3}{16} \times 18 = 3\frac{3}{8}$ miles
Second turn to Hwy 123: 3/4"	$\frac{3}{4} \times 18 = 13\frac{1}{2}$ miles
Hwy 123 to Price: 7/8"	$\frac{7}{8} \times 18 = 15\frac{3}{4}$ miles

Hands On, Inc
2121 Rebild Drive
Solvang, CA 93463

							millions
							hundred thousands
							ten thousands
			Free				thousands
							hundreds
							tens
							ones

HUNDRED BILLIONS	TEN BILLIONS	BILLIONS	HUNDRED MILLIONS	TEN MILLIONS	MILLIONS	HUNDRED THOUSANDS	TEN THOUSANDS	THOUSANDS	HUNDREDS	TENS	ONES

Making a Spinner

Many activities in this book require the use of spinners. Students can make their own spinners very easily and inexpensively by following the directions given here.

Materials required are: cardboard, scissors, compasses, rulers, pencils, paper clips, a paper punch, and tape.

1. Cut out a cardboard (tagboard) pointer and punch a hole in the center.

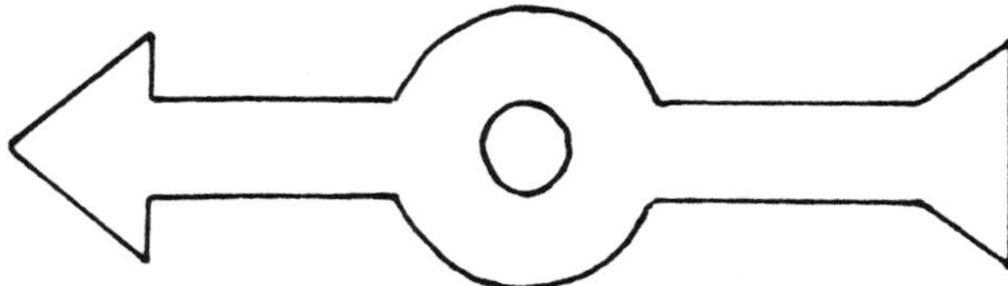

2. Cut a scrap of cardboard as a paper washer and punch a hole in the center.

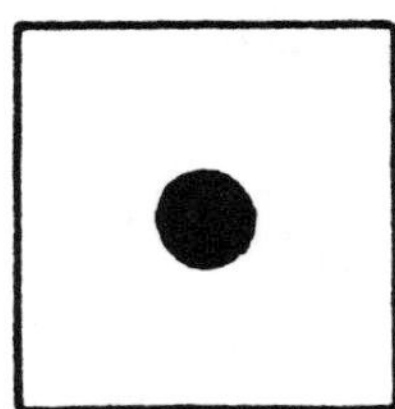

3. Cut out a four inch square of cardboard and divide it into quarters as shown. Make light pencil lines.

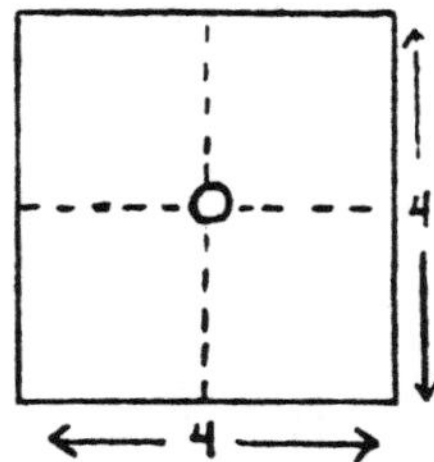

4. At the center of the square, make a small hole with your paper clip.

5. Using a compass, make a circle on the four inch square and draw a design you wish to use. You may wish to color your spinner at this point.

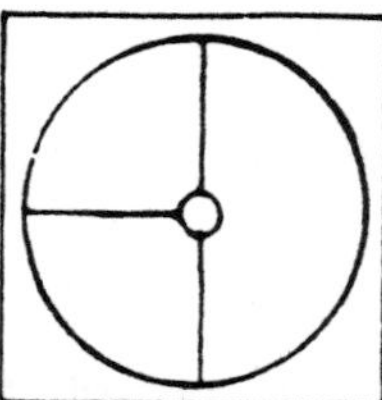

6. Bend the center loop of the paper clip up at a 90 degree angle to the outer loop.

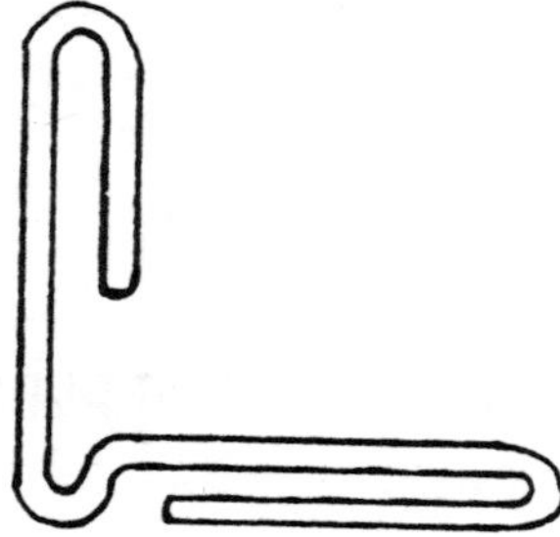

7. Tape the outer loop of the paper clip to the bottom of the four inch square to hold it in place.

8. Put the bent paper clip through the hole in the four inch square, the paper washer and the pointer.

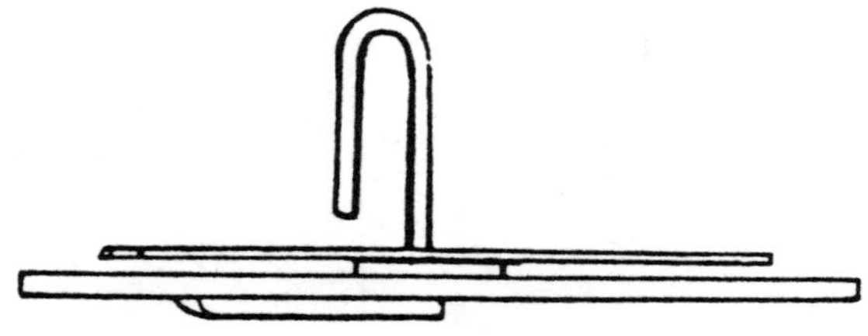

1	2	3	4	5	6	7	8	9	10
11	12	13	14	15	16	17	18	19	20
21	22	23	24	25	26	27	28	29	30
31	32	33	34	35	36	37	38	39	40
41	42	43	44	45	46	47	48	49	50
51	52	53	54	55	56	57	58	59	60
61	62	63	64	65	66	67	68	69	70
71	72	73	74	75	76	77	78	79	80
81	82	83	84	85	86	87	88	89	90
91	92	93	94	95	96	97	98	99	100

$\frac{1}{2}$	$\frac{5}{10}$	$\frac{3}{4}$	$\frac{6}{8}$
$\frac{1}{3}$	$\frac{3}{9}$	$\frac{1}{4}$	$\frac{3}{12}$
$\frac{2}{5}$	$\frac{4}{10}$	$\frac{3}{5}$	$\frac{9}{15}$
$\frac{2}{3}$	$\frac{4}{6}$	$\frac{4}{5}$	$\frac{8}{10}$

$\frac{1}{6}$	$\frac{2}{12}$	$\frac{3}{8}$	$\frac{6}{16}$
$\frac{5}{6}$	$\frac{15}{18}$	$\frac{1}{5}$	$\frac{3}{15}$
$\frac{1}{7}$	$\frac{2}{14}$	$\frac{5}{8}$	$\frac{15}{24}$
$\frac{1}{8}$	$\frac{2}{16}$	$\frac{7}{8}$	$\frac{21}{24}$

$\frac{1}{9}$	$\frac{2}{18}$	$\frac{3}{10}$	$\frac{9}{30}$
$\frac{2}{9}$	$\frac{6}{27}$	$\frac{7}{10}$	$\frac{14}{20}$
$\frac{5}{9}$	$\frac{10}{18}$	$\frac{1}{12}$	$\frac{2}{24}$
$\frac{1}{10}$	$\frac{2}{20}$	$\frac{11}{12}$	$\frac{22}{24}$

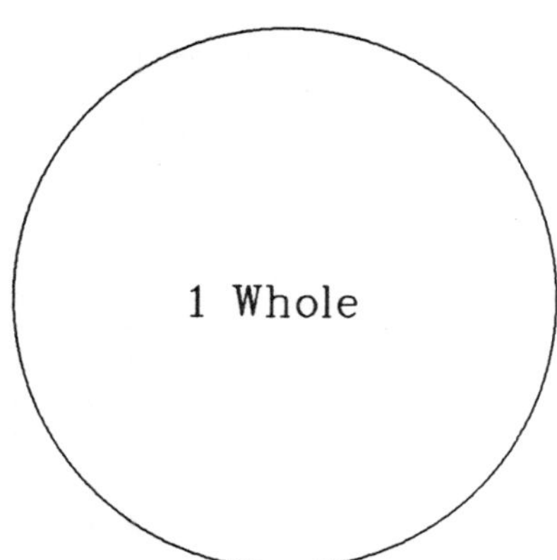
1 Whole

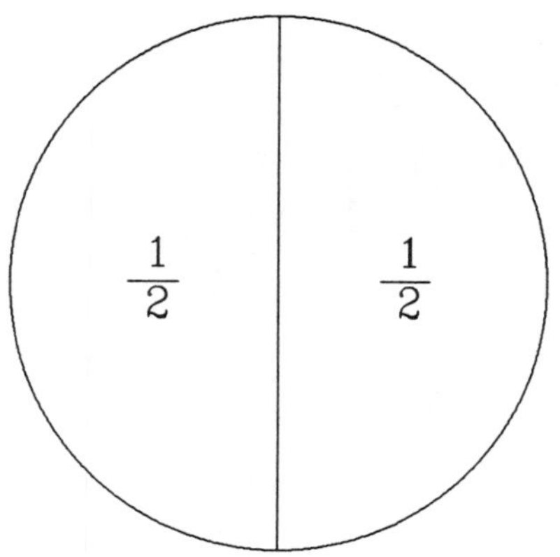
1/2
1/2

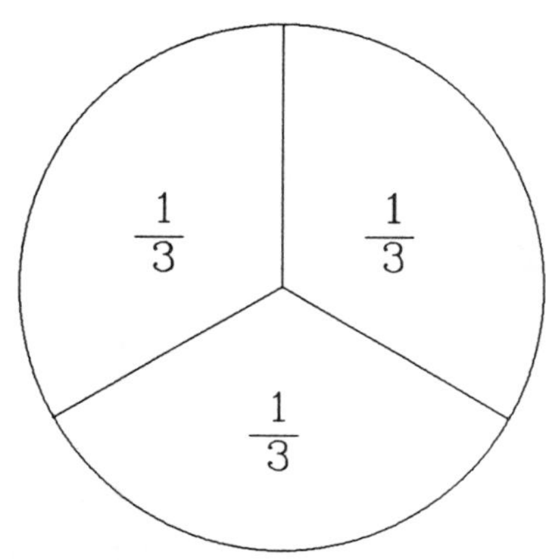
1/4
1/4
1/4
1/4

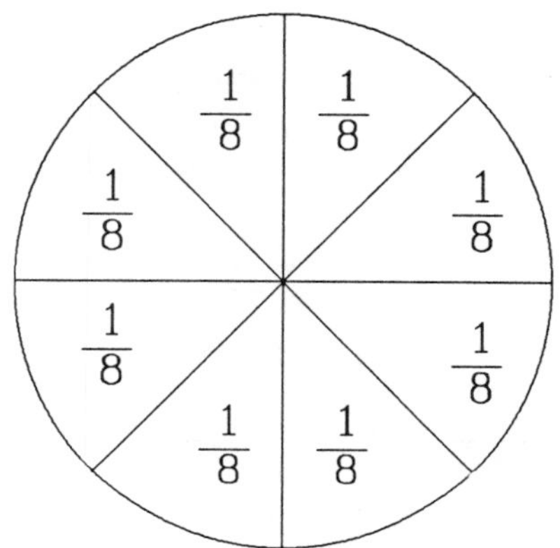
1/8
1/8
1/8
1/8
1/8
1/8
1/8
1/8

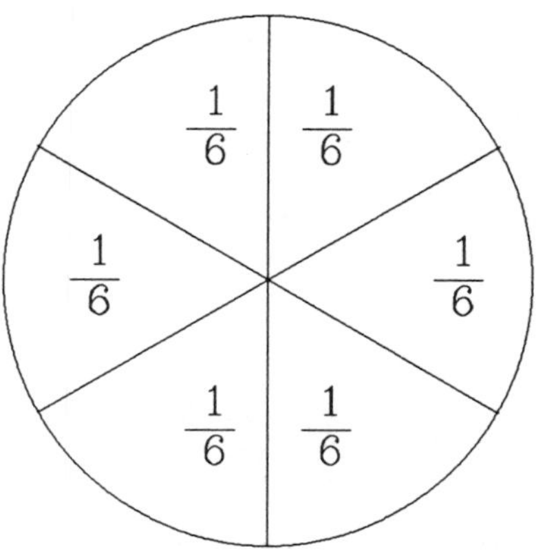
1/3
1/3
1/3

1/6
1/6
1/6
1/6
1/6
1/6

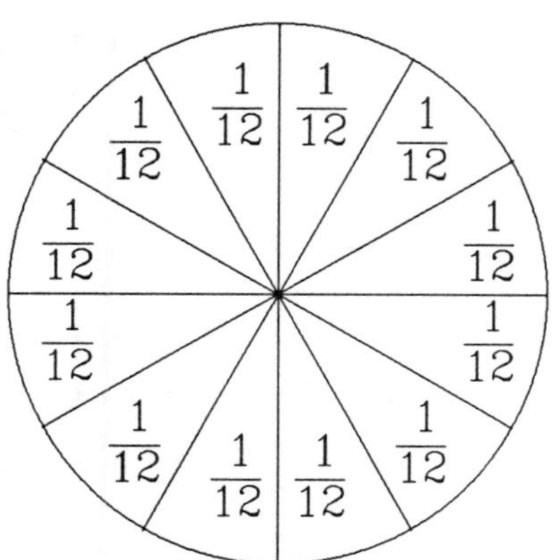
1/12
1/12
1/12
1/12
1/12
1/12
1/12
1/12
1/12
1/12
1/12
1/12

$\frac{5}{10}$	$\frac{4}{10}$	$\frac{3}{10}$
$\frac{2}{10}$	$\frac{1}{10}$	$\frac{1}{10}$
$\frac{5}{10}$	$\frac{4}{10}$	$\frac{3}{10}$
$\frac{2}{10}$	$\frac{1}{10}$	$\frac{1}{10}$
$\frac{1}{20}$	$\frac{2}{20}$	$\frac{3}{20}$
$\frac{1}{20}$	$\frac{2}{20}$	$\frac{3}{20}$
$\frac{2}{25}$	$\frac{1}{25}$	$\frac{1}{25}$
$\frac{2}{25}$	$\frac{1}{15}$	$\frac{1}{15}$
$\frac{1}{2}$	$\frac{2}{15}$	$\frac{2}{15}$

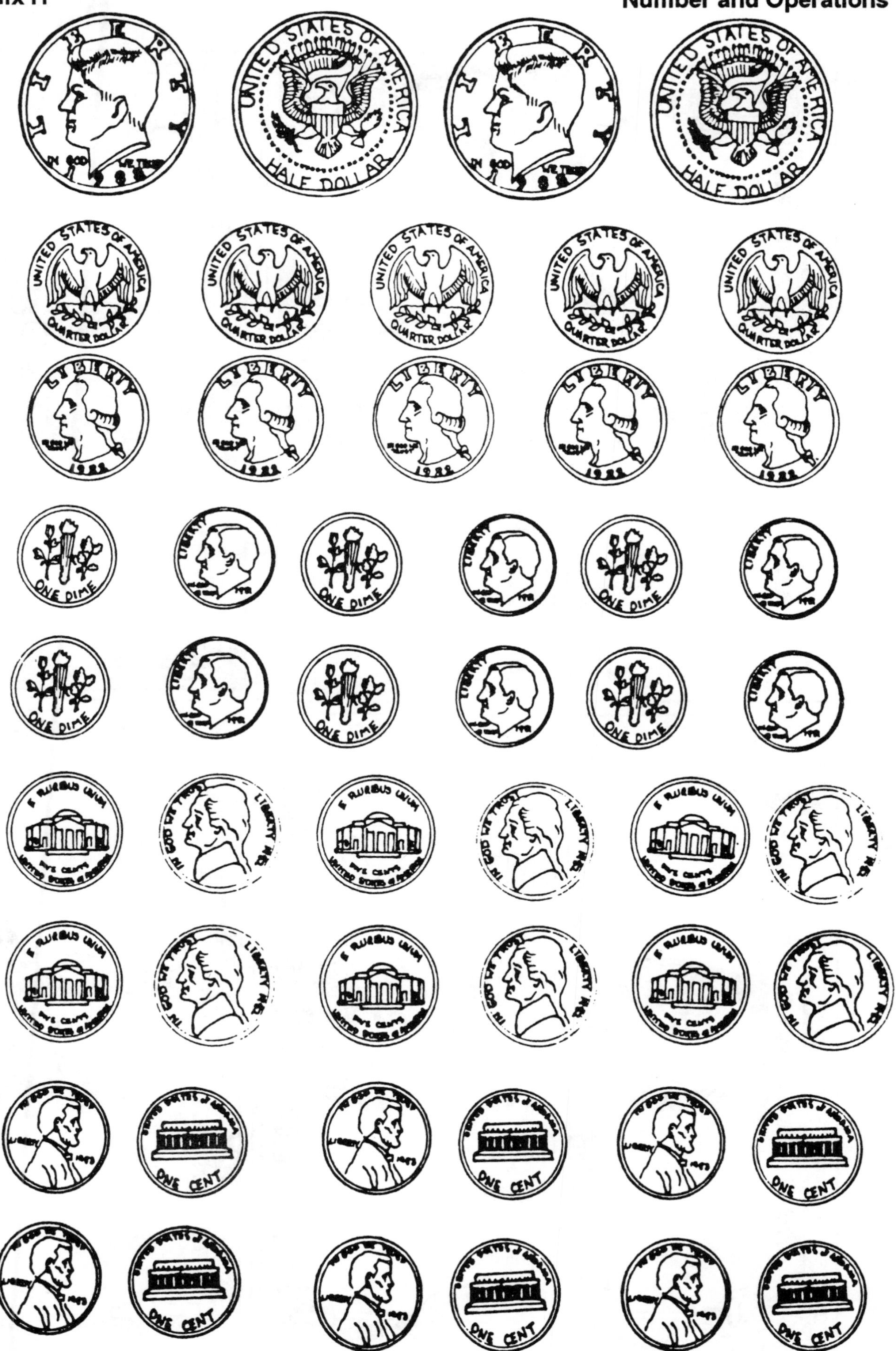

FEDERAL RESERVE NOTE
THE UNITED STATES OF AMERICA
ONE DOLLAR